71+10
Magic Tricks
for Children

V&S PUBLISHERS

Published by:

V&S PUBLISHERS

F-2/16, Ansari Road, Daryaganj, New Delhi-110002
011-23240026, 011-23240027 • *Fax:* 011-23240028
Email: info@vspublishers.com • *Website:* www.vspublishers.com

Branch : Hyderabad
5-1-707/1, Brij Bhawan (Beside Central Bank of India Lane)
Bank Street, Koti, Hyderabad - 500 095
040-24737290
E-mail: vspublishershyd@gmail.com

Follow us on:

For any assistance sms **VSPUB** to **56161**

All books available at **www.vspublishers.com**

© **Copyright:** *V&S* PUBLISHERS
ISBN 978-93-505702-1-0
Edition 2013

Printed at : Param Offsetters, Okhla, New Delhi-110020

Publisher's Note

V&S Publishers is glad to inform its esteemed readers that we have come up with this new and exciting book on Magic that contains some of the most fascinating Magic Tricks. This has been put under the **'71 Series'** which includes some of our best-sellers, like *71 Science Experiments, 71 + 10 New Science Projects* and there are many more such interesting books under printing. Actually, all these books in the '71 Series' have been in popular demand for almost two decades.

The present book is not just a collection of **'71+10'** fully coloured amazing Magic Tricks with the secret behind performing each trick, but also has a brief compilation of the life histories of some of the best and world famous magicians from across the globe.

The book is definitely a must read for all those enthusiastic readers who want to read, learn and master the Art of Magic and of course, the children – who are always eager and thrilled by the word, **'Magic'**. All the tricks described in the book have been written in a simple and comprehensive language with attractive pictures that makes it all the more fascinating for our young readers, particularly the school children of all age groups.

Though every effort has been made to minimise printing and other proofreading errors, it could be that a few might have managed to escape our wakeful eyes. We would like to request our enlightened readers to bring these errors to our notice so that we can rectify the same in subsequent editions. Hope you enjoy reading the book and practising some of the tricks too!

Contents

Getting Started in the Art of Magic

Since the beginning of time humanity has been fascinated with strange occurrences, the paranormal and anything considered "magical." Magic has been the inspiration and plot for many stories throughout time, and it never gets old. Everyone wants to believe that there is something else there; therefore, they love to watch those who seem to exhibit proof. What if you could be the person who offers that proof? That is the art of magic of a magician.

Magic is a very rewarding hobby, and not just in terms of financial gain. It benefits everyone who takes part–both the audience and the magician. It can take those who watch back to a childlike state of wonder and amazement and make them forget daily troubles. It allows them to go into another world, a world where magic indeed does exist, and as far as entertainment arts and hobbies go, magic is one of the few that accomplishes this. Magic can also break down language barriers and cross cultural lines. There are so many good things about it that I could go on forever, but you are probably waiting to find out how does one get started in this wonderful hobby, and what does it require?

If you are interested in magic, the first thing any practicing magician will tell you is to read some books on the subject, and I would agree with that. First head to the library and check out as many books as you can find. Don't worry if the books seem childlike or for younger kids; you will most likely still find some great things for beginners, and many people tend to overlook those kinds of books, way too often. You will want to learn some of the basic tricks from those kinds of books before moving on. Every magician starts out this way, and it works.

Once you learn some of the basics and have found some tricks you like, the first thing you will want to do is to perform them right away but DON'T. This is a mistake that many aspiring magicians make. This method overlooks one of the parts of learning that a lot of children or beginners overlook, and that is practice. Practice really is key to a magician's performance. None of the greats reached their point of expertise overnight, but by many days of practice. However, practice does NOT have to be boring; it can be fun. If you have friends or family willing to help, practise with them. If another person isn't available, a full-length mirror works about as well. Some tricks may seem difficult at first, but just remember to keep trying until you get it down smoothly and until you can perform it, while, talking and doing other things at the same time.

Then comes the performances. At first you may just perform the magic in front of family and friends, but your confidence will grow (as you practise and perform more), and soon you will be comfortable performing before anyone anytime. So keep practising and performing and trying tricks that work for you and that fit your personality.

Many of you who are reading this have tried this approach, perhaps, when you were younger, or maybe even recently. And then you were lost. You asked, "Well, what now?" And there was never any easy answer. This is the reason many go through just the magic "phase" and get interested, but then stop and forget it. You didn't know what to do or how to advance from that point. Really, though, you just have to look around as there is knowledge everywhere on how to continue. Search the internet, find a local magic shop anywhere. You can find tons of videos on magic online, and there are many great stores. However, the point is that the knowledge is out there, and once you get some basics down, then it is time to seek that knowledge.

It is imperative that you have a curiosity and a thirst for that magical knowledge, and you may not believe it now, but as you learn more, you will eventually begin creating your own tricks. That's right; your very own, that you make all by yourself that is a very rewarding experience, and you CAN do it. Magic, like ANY other hobby, mainly just takes dedication. If you have a dedication to what you do, you can do almost anything you want. It takes a dedicated mind to practise and perform to boost your confidence. The best part of it all will come in the audience's reactions; that is your real reward. When you see that person's expression of amazement, of joy that is a priceless moment and makes all the practice worthwhile.

This in short is the art of magic. There is so much more to it, but this is an overview. If it sounds simple, so it should. It just takes a keen mind and a strong desire for knowledge. So if this interests you in the slightest, go down to your local library or magic shop, and start learning! It really is so much fun that once you get started, and I think you will love it as so many have before!

New to Magic?

If you are new to magic and want to take up conjuring as a hobby or even a future career, the very best way to begin is to watch as much magic as you can by regularly attending magic shows, watching television on magic programmes carefully and so on.....

This will help in many ways such as:

You can see what type of magic appeals to you, whether it is a *close-up magic like cards* and *coins*, or even stage illusions. This will help you to decide which avenue of magic you wish to start with. Whilst magic can be a relatively cheap hobby, if you begin with learning cards or coins, it does take a large investment of time to practise and perfect your skills, so make sure you are committed to spending a proportion of your free time on something that you are really motivated to learning.

Also, by watching as much magic as possible, whether live or on TV, etc. you get to see many different ways of presenting your magic. Although most hobbyists new to magic naturally want to learn as many new tricks as possible, you will soon see that it is the presentation of your effects that will gain you the status of a 'magician' and not the mechanics of the trick.

When you have seen scores of magic, you get a feeling of what type of material would suit your personality. If you can combine a convincing presentation of your effects that fits your personality, you will be one step ahead of the vast majority of hobbyists that like to call themselves 'magicians'!

Finally, try to narrow your choice of effects (tricks) to around 5 or 6 initially, which you can competently perform and work on improving them until you feel confident enough to show them to the people. The feedback you get will be the most valuable advice that anyone can ever give you about magic, but prepare yourself for some of it to be unflattering or even brutal at first! This will help you in selecting and rejecting material until you have a core 'repertoire' of pieces that you are happy with. Perform your magic for strangers; because they don't need to pander to your ego and will probably be the most honest in giving their opinion. If you can build on the good feedback and eliminate the problems that they highlight, you become worthy to call yourself a 'MAGICIAN'.

Good luck, and enjoy the ride!

Basic Tips for Beginners

Many magic tricks and effects are fairly easy to learn. The toughest part is to actually make the audience believe they have just seen something magical in front of their eyes. The things that need to be learnt in order to pull this off is *misdirection* and above all, *confidence*. If you're not confident while performing a trick, the audience will see through this and will lose interest almost immediately.

One of the most used items in magic and also one of the oldest is the *playing cards*. This makes it one of the best starting points for any budding magician. There are special gimmick cards that can be purchased to make the tricks even easier but it is highly recommended to learn the tricks with normal cards. This way, the whole deck can be examined before and after the trick.

Practice

You may have heard the term, "Practice, Practice, Practice". This is no more true than it is in magic. When I started I practised every chance I got to master the art of performing tricks. Plenty of practise does not only allow you to master the trick but also build confidence in what you're doing. This in return makes it more unlikely for anything to go wrong when performing in front of other people or a large audience.

During the early stages, it is important that you don't try to jump straight in to the difficult tricks. You MUST first take the time to learn the *basics* and this includes shuffles, sleights and easy tricks. Each trick that then follows will come much easier and you'll be able to pick them up much more quickly. Many *props* can be easily found around the house. *Cups, balls, cards and coins.* Begin with these items before going out and spending money on props. This way you'll know if magic really is for you, or you are made for magic!

Misdirection

The next step is to think about the audience. You need to make sure that they do not see how the trick is done. This is called *distraction* and *misdirection*. If you're trying to conceal something in your hand then you don't want them to look directly at it. Different magicians have different ways of doing this. Some use humour whilst others will just simply ask the audience questions. This way the attention is off of you for long enough to do what you need to do.

Lastly, come up with your own style. Don't copy any other magician. People like something original which they have not seen before. Everybody has their favourite magicians, but this does not mean you have to copy everything about them. You'll also get much more respect as a magician by creating something new and different.

Interesting Facts about Magicians

The word 'magic' is derived from the Persian word 'magus' which designated a priestly class.

Magic has many names! It is also called conjuring, hocus pocus, sorcery and wizardry, to name some of the most common ones.

The most dangerous trick in magic is the Bullet Catch. This effect, in which a marked bullet is fired at the performer who catches it between his teeth, has killed twelve magicians and wounded many more.

Harry Houdini died on Halloween in 1926. This brilliant magician and escape artist was the first man to fly an airplane in Australia on March 16, 1919.

The great film director, Orson Wells had a lifelong interest in magic. During World War II, he had his own magic show that he presented for members of the U.S. armed forces. His assistants at times included such stars as Rita Hayworth and Marlene Dietrich.

Matthew Buchinger, one of the premier Cups and Balls performers of the 18th century, was born without arms or legs and was 29 inches tall.

Charles Dickens was an enthusiastic amateur magician. In August 1849, in one of his most ambitious performances he introduced himself as "The Unparalleled Necromancer Rhia Rhama Rhoos, educated cabalistic ally in the orange groves of Salamanca and the ocean caves of Alum Bay."

David Copperfield is the first living magician to have a star on the famous Hollywood Walk of Fame. The only other magician so honoured is Harry Houdini, who received a star after his death.

The Levitation Illusion was first performed in Greek dramas as early as 431 B.C.

The ancient Greeks were great admirers of magic, erecting statues of their favourite magicians. Homer even mentions conjurors in his epic poem, The lliad.

The most famous Chinese magician of all time, Chung Ling Soo, was really an American named William E. Robinson. He was mortally wounded in 1918 doing the Bullet Catch trick on the stage of the Wood Green Empire Theatre in London and died the next day. Only then did the world discover that he was not Chinese.

During World War II, the magician Jasper Maskelyne hid the Suez Canal and Alexandria Harbour from the Germans and helped the Allied Forces win the war in Africa. In the book Top Secret, Maskelyne tells of his war experiences and of the time when he performed at the Empire Theatre in Cairo, Egypt as "The Royal Command Magician." Few people actually realized that the performance was a front for the British intelligence Service.

Magicians were very much involved in the birth of the movie industry. Not only were many magicians exhibitors of films, but many were involved as performers and producers. Harry Houdini made several silent films and was the creator of many special effects; magician George Melies bought the Robert-Houdin Theatre and exhibited the first motion picture seen in Paris.

The author of the 14 most recent James Bond thrillers is a magician. John Gardner, retained by the estate of Ian Fleming, the creator of the Bond character, was a professional magician before he became an author.

David Copperfield is the highest paid magician being named on fortune 500's list.

The world's fastest magician is Eldon D. Wigton (Dr. Eldoonie). He performed 255 tricks in 2 minutes on April 21, 1991.

Eliaser Bamberg, the 18th-century Dutch magician, was known as "The Crippled Devil." He had lost one of his legs in an explosion and wore a wooden leg. The story goes that Eliase had hollowed out his wooden leg and used it as a secret hiding place for his magic props.

The worlds strongest magician is Ken Simmons, He can bench press over 500lbs

Famous Celebrities who are (were) also Magicians: Johnny Carson, Don Johnson, Woody Allen, Dick Cavett, Dick Van Dyke, Milton Berle, Cary Grant, Bill Bixby, Jimmy Stewart, Steve Martin, Muhammad Ali, Bob Barker, George Bush, Jerry Lewis, Charles Dickens.

David Copperfield

David Copperfield (born as David Seth Kotkin on September 16, 1956) is an American illusionist, and has been described by Forbes as the most commercially successful magician in history.

Copperfield's television specials have won 21 Emmy Awards out of a total of 38 nominations. Best known for his combination of storytelling and illusion, Copperfield's career of over 30 years has earned him 11 Guinness World Records, a star on the Hollywood Walk of Fame, and a knighthood by the French government; and he has been named a Living Legend by the US Library of Congress.

Copperfield has so far sold 40 million tickets and grossed over $3 billion, which is more than any other solo entertainer in history. He currently performs over 500 shows a year.

When not performing, he creates experiences for his chain of eleven islands in the Bahamas – Musha Cay and the Islands of Copperfield Bay, which has completed a $35 million renovation under Copperfield's supervision.

When Copperfield was 10, he began practising magic as "Davino the Boy Magician" in his neighbourhood, and at the age of 12, became the youngest person ever admitted to the Society of American Magicians. Shy and a loner, the young Copperfield saw magic as a way of fitting in and, later, as a way to get girls. By the age of 16, he was teaching a course in magic at the New York University.

At the age of 18, Copperfield enrolled at Fordham University and was cast in the lead role of the Chicago-based musical The Magic Man three weeks into his

freshman year, adopted his new stage name, 'David Copperfield' from the famous Charles Dickens novel. At the age of 19, he was headlining at the Pagoda Hotel in Honolulu, Hawaii.

Copperfield's career in television began in earnest when he was discovered by Joseph Cates, a producer of Broadway shows and television specials. Cates produced a magic special in 1977 on ABC called "The Magic of ABC" hosted by Copperfield, as well as several of "The Magic of David Copperfield" specials on CBS between 1978 and 1998. There have been 20 Copperfield TV specials between 1977 and 2001.

Most of his media appearances have been through television specials and guest spots on television programmes. His illusions have included making the Statue of Liberty disappear, flying, levitating over the Grand Canyon, and walking through the Great Wall of China.

In 1996, in collaboration with Francis Ford Coppola, David Ives, and Eiko Ishioka, Copperfield's Broadway show Dreams & Nightmares broke box office records in New York at the Martin Beck Theatre. Reviewer Greg Evans, described the sold-out show in Variety magazine: "With a likable, self-effacing demeanour that rarely comes across in his TV specials, Copperfield leads the audience through nearly two hours of truly mind-boggling illusions. He disappears and reappears, gets cut in half, makes audience members vanish and others levitate. Copperfield climaxes his show with a flying routine, seven years in the making, that defies both logic and visual evidence -- he could probably retire just by selling his secrets to future productions of Peter Pan."

Also during 1996, Copperfield joined forces with Dean Koontz, Joyce Carol Oates, Ray Bradbury and others for David Copperfield's Tales of the Impossible, an anthology of original fiction set in the world of magic and illusion. A second volume was later published in 1997, called David Copperfield's Beyond Imagination.

On May 7, 2009, Copperfield was dropped by Michael Jackson from Jackson's residency at the O2 Arena after an alleged row over money. Copperfield wanted $1 million (£666,000) per show. Copperfield denied the reports of a row, saying "don't believe everything you read." News of Copperfield's collaboration with Jackson first surfaced on April 1, 2009, and has since been reported by several websites as a possible April Fool's prank.

In August 2009, Copperfield brought his show to Australia.

In January 2011 David Copperfield joined the cast of the new feature film Burt Wonderstone with Steve Carell, Jim Carrey, James Gandolfini, and Olivia Wilde. Copperfield and his team also developed illusions used in the film.

In July 2012, OWN-TV network aired a one-hour special interview with Copperfield as part of the network's "Oprah's Next Chapter" series. The show featured many aspects of Copperfield's personal life and family -- with tours of his island home and Las Vegas conjuring museum -- and a sampling of his illusions and magic effects. During the interview, he and his girlfriend Chloe Gosselin, a French fashion model, announced their engagement, and appeared together briefly with their young daughter strolling down the beach on the island.

1

The Power of Wand

Effect: The magician makes some mysterious passes around a wand or pencil which uncannily begins to move on its own.

Secret: The magician secretly blows on the wand, which causes it to roll.

Props:

Preparation: Practise blowing towards the wand gently and secretly.

Presentation: Lay the wand on the table and very slowly trace circles around the

blowing towards the wand, which causes it to roll easily on the smooth surface.

2

Ring and String

Effect: A ring threaded onto a loop is released, although a spectator is holding the string.

Secret: Read the preparation carefully to see how you can solve this puzzle.

Props: Any ring and one cord.

Preparation: Get someone to help you practise this trick before presenting it.

the cord. As he/she does so, you let go of your loop. The ring will fall free even

of times before you can expect to do it smoothly.

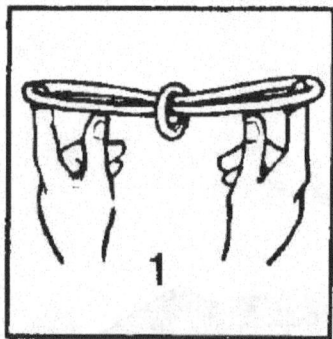 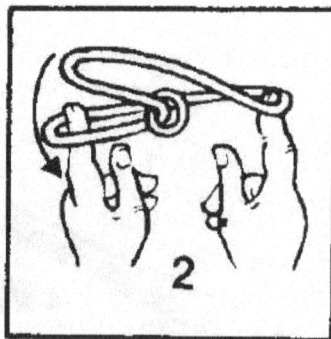 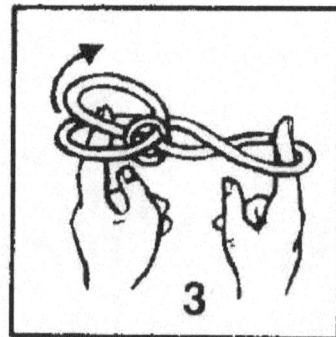

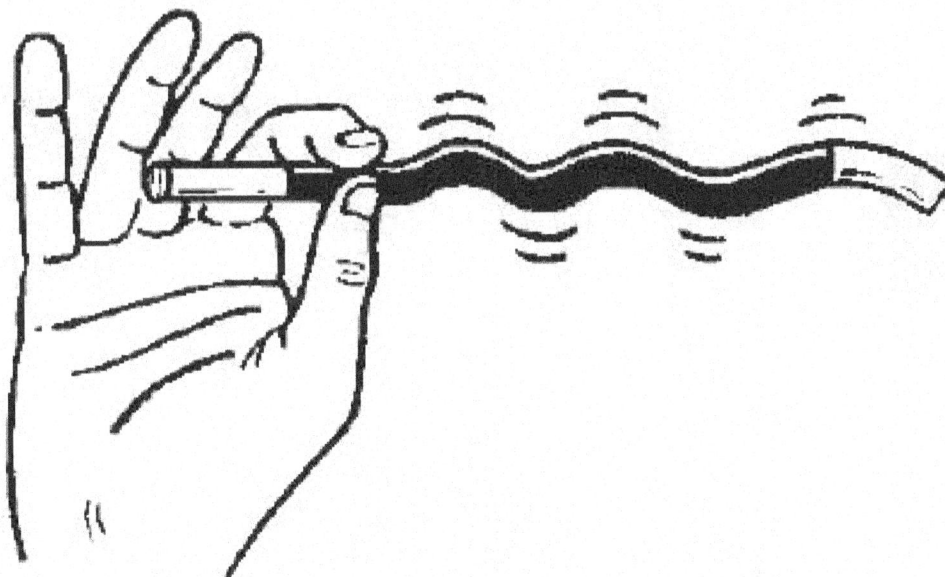

An Elastic Wand

Effect: The magician's solid wand or pencil seems to turn to soft rubber. On command, it turns solid again.

Secret: Hold the wand horizontally in front of you between your thumb and

optical illusion.

Presentation: Tap the wand on something to show that it is solid. Say the magic

and it turns solid again. Interesting, isn't it?

4

Palm the Wand

Effect: To prove that you have magical powers, place the wand or pencil across the top of the palm of one hand, with the back of the hand towards the audience. Slowly

Secret:

see the way you hold the wand. This is again an *optical illusion*, and of course, a simple yet amzing trick!

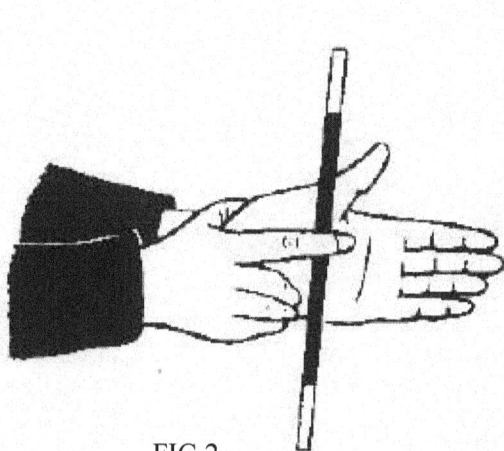

FIG 2

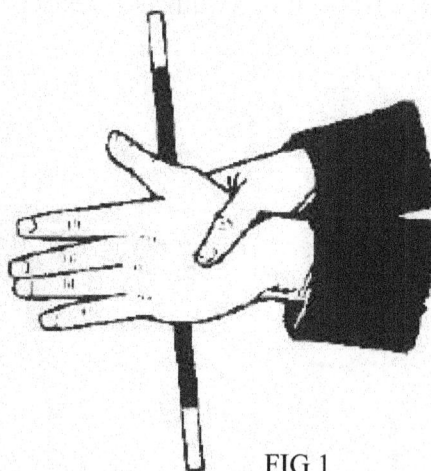

FIG 1

Reverse Loops

Effect: A cord, coiled around a magic wand or pencil and tied through a ring, is pulled free of the wand and yet the ring remains securely tied.

Secret: As you wind on the cord, you secretly reverse the direction.

Props: A magic wand or pencil, a ring and a length of cord.

Preparation: Practise coiling the cord around the wand until you can do it without hesitation.

Presentation: Take the cord and wind it around the magic wand in one direction

Practise this carefully and try to get the ends of the cords to hang down about the same lengths. Cover all of the loops with your free hand. Now, have someone tie a ring to the two loose ends of the cord. As you say a mysterious magic word, have him/her pull straight down and the cord will come free from the wand.

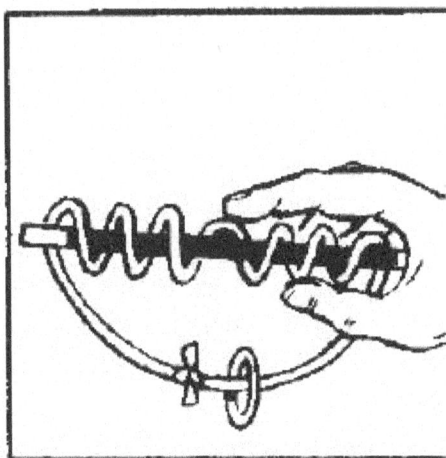

Harry Houdini

Harry Houdini is still regarded as one of the greatest magicians and escape artists of all time. Born as Ehrich Weiss in Budapest, Hungary, he and his family emigrated to the United States, where he was soon to be called "Harry" by his friends.

After his father took him to see a travelling magician, Harry Houdini's interest in magic and performing arts soon appeared at the forefront of his mind.

At 15, he read an autobiography by Jean-Eugene Robert-Houdin, who was, at the time, the father of modern magic. This was the inspiration behind his name change to Houdini - one of his friends said that by adding an "i" to Houdin, it meant "like Houdin". This is how Harry Houdini was born.

He initially persisted with traditional card and coin magic, where he called himself the "King of Cards". One trick was the "card on the ceiling". This consists of a card selected by a spectator, signed and returned to the deck. Harry Houdini would then throw the shuffled pack at the ceiling and the spectator's card would stick to the ceiling.

Card tricks were easy for him, so he soon began trying escape acts. Harry Houdini's major breakthrough came when he met another showman, Martin Beck, who was so impressed with Houdini's handcuffs act that he advised him to concentrate on escape acts.

Early in his career, Harry Houdini met another performer who was to later become his wife, Wilhelmina Beatrice Houdini. They performed together as The

Houdinis often performing the Metamorphosis illusion. In this illusion, Houdini would be locked inside a large chest, usually with handcuffs and ropes. His wife is atop the trunk, briefly holding a curtain concealing her body. The curtain is lowered and Houdini is now on top of the chest with his wife, restrained as Houdini had been, inside the chest.

He travelled to Europe where he became a star. Harry Houdini would free himself from strait-jackets, chains, ropes, handcuffs and even jails.

When he returned to the States, he introduced ever more dangerous acts. One of which was the Giant Milk Can Escape. This is where Harry Houdini escaped from inside a padlocked water-filled milk can.

Sadly, Harry Houdini died at the age of 52, from a ruptured appendix.

He had also left his wife a secret code that would be used by him, to contact her, after his death, should he have been wrong about spiritualists. For ten years, his wife held a séance once a year, but Harry Houdini never made contact.

6

A Vanishing Beaker

Effect: A beaker vanishes.

Secret: The beaker is glued to a small wooden tray. In the centre of a scarf, sew a circle of cardboard. Show the beaker and then cover it with the scarf or handkerchief.

on your table – with the beaker side away from the audience. Everyone believes you are holding the beaker. Throw the scarf in the air and catch it as it falls – the beaker has vanished.

CUP STUCK TO
TRAY WITH A
DOUBLE SIDED
TAPE

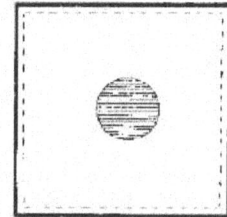

CARDBOARD STUCK TO A
LARGE HANDKERCHIERF

Ring On

Effect: A ring is magically threaded on a string.

Secret: Place the string and ring on a table, together with a safety pin and a handkerchief. Cover the string, ring and pin with the handkerchief. Allow the ends of the string to remain visible at all times. Place your hands beneath the handkerchief and push the centre of the string through the ring. Open the pin and then close it so that it goes around the right side of the loop and the rest of the string through the ring. Open the pin then close it so it goes around the right side of the loop and the

Remove your right hand and lift up the right end of the string and ask someone to

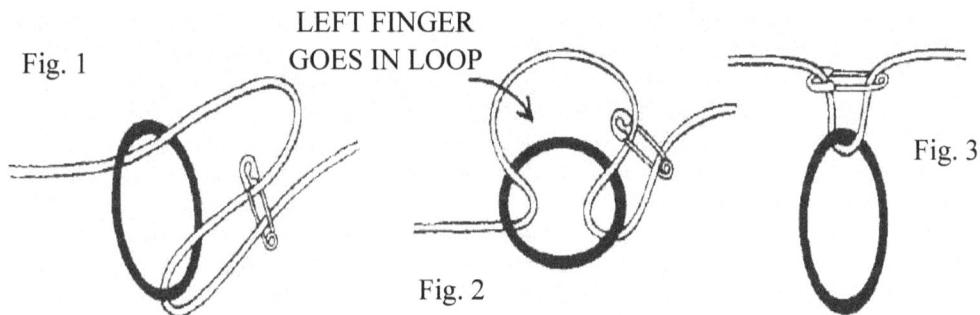

Fig. 1

LEFT FINGER
GOES IN LOOP

Fig. 3

Fig. 2

A Floating Beaker

Effect:

Secret: On one side of a plastic beaker attach a strip of strong paper. It is important

the air, you push your right thumb through the paper loop. Place your left hand near your right hand and lift your hands in the air. With practice, it will appear that the

freely without falling down. Great, isn't it?

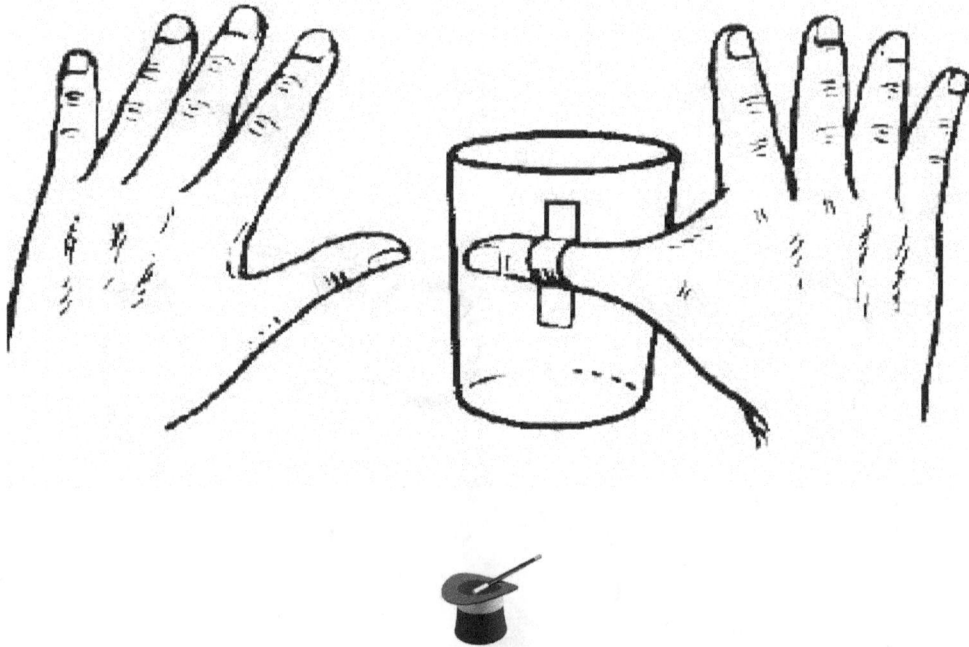

The Removable Thumb

Effect: You appear to remove your thumb!

Secrets: Practise this in front of a mirror. Hold out your left hand with the palm facing you. Bend your left thumb in half towards your body. Bend your right thumb

done only to people who are directly facing you.

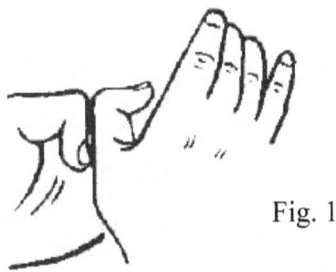

Fig. 1

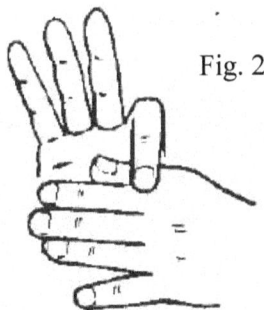

Fig. 2

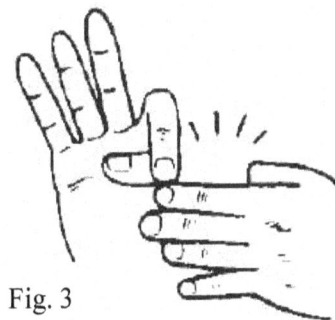

Fig. 3

Folding Money

Effect: A currency note turns itself over.

Secret: Hold the note in your left hand, then fold it in half lengthways towards you. Next fold it in half away from you. Finally, fold the note in half again, this time towards your body. Now slowly unfold the note from the front, making each movement in the same direction. The note is now upside down! Strangely enough,

practise it thoroughly to make sure you can do it perfectly before showing it to anyone.

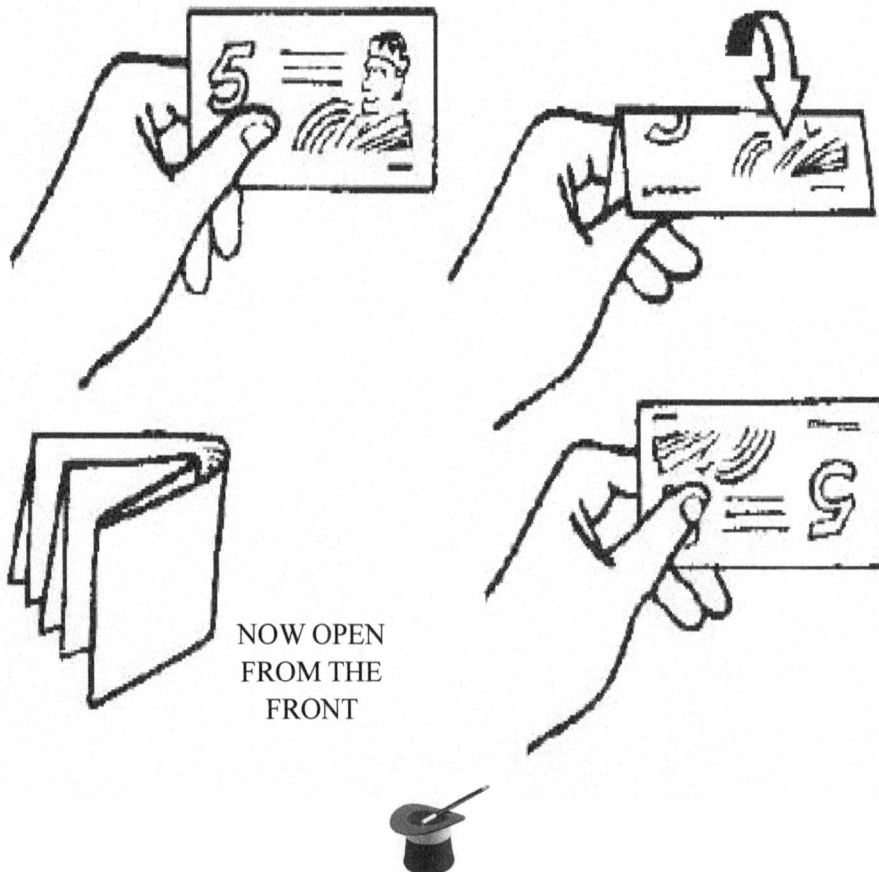

NOW OPEN
FROM THE
FRONT

An Obeying Matchbox

Effect: A matchbox obeys your commands.

Secret:

Then close the box. You are now ready to show the tricks. Hold each end of the thread, with one hand above the other. The matchbox should be near the top of the thread. If you hold the thread loosely, the box will slide down the thread but if you

make it stop or go at your command.

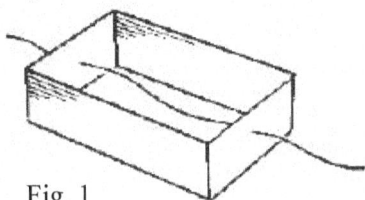

Fig. 1

COTTON
THROUGH
DRAWER

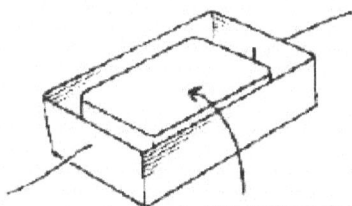

Fig. 2

ERASER IN
DRAWER

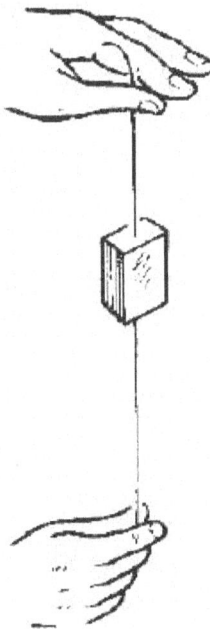

STOP!

Fig. 3

Pir Mohammad Chhel

Mohammed Chhel originally a fakir (mystic) was a renowned magician of Gujarat. Mohammed Chhel was born in 1850 in Ningala, a small village of Bhavnagar district. Basically he was a Pir of a known Dargah and was involved his whole life in benevolent works. Said to have blessed with supernatural powers, Mohammed Chhel eventually turned magician but his character and nobility, message of life he wanted to convey to people, himself serving the society and helping needy with his magnetic aura and some miraculous acts flavoured by his fine sense of humour remained more a qualities and characteristics of a mystic.

However, he does not fit into the usual mould of sleight-of-hand artistes or tricksters like other great magicians. Mohammed Chhel had an impressive extrovert personality.

With his metal-rimmed glasses, he looked more of an intellectual. He had a prominent nose, thin lips, flowing beard and slightly upturned moustaches. A coiled turban and a waistcoat made him look urbane.

In the closing years of 19th century and the early part of the 20th, Mohammed Chhel had become quite famous. Chhel was not a stage performer and had no use for any props or equipments. He performed impromptu against the backdrop of spontaneous flow of life.

He differed from the other magicians in the way that he was an instant conjurer and most of his acts were performed on running trains that traveled to and fro the countryside. The passengers were peasants and simple village folk. He entertained them with his acts, occasionally extending support to the needy.

His performances were meant to define life more emphatically. There was no cause and effect relationship in his acts and they transcended material bounds of reality.

He could make a train ticket checker shell-shocked by producing an avalanche of tickets out of his chin. He could decouple a running train with only the engine chugging away.

He freed a poor peasant from the clutches of moneylender by casting a spell that will not allow him to get out of his chair. Only when he wrote off the falsified debts, did Chhel let the errant moneylender go unstuck.

A troubled merchant wanted to know about his ailing wife in Mumbai. Those were the days when telephones were not a common facility for the village folk. So he came to Mohammed Chhel who put his palm across the merchant's eyes and asked him if he could see his wife. The merchant was relieved to see his wife working in the kitchen hale and hearty. Later the news of her well-being reached him, confirming the powers of Mohammed Chhel.

Mohammed Chhel was not a magician in the usual sense of the term. He made things happen for the benefit of fellow creatures. The date of his death is not known but probably he lived till, early twenties of the last century. People still remember his nobility.

Kings and Aces

Effect: Two playing cards make a mysterious journey.

Secret: You need special playing cards. First glue an Ace of Spades and a King of Diamonds back to back. Cut an Ace of the Clubs in half lengthways. Glue this on top of the Ace of Spades. Now glue a smaller strip of another King of Diamonds on top of that. Hold a King of Hearts against the special card and it looks like four cards. Place the four cards in a box. Bring out the King of Hearts and place it in another box. Secretly turn the special card over and bring it out, showing the

showing them as you did before. It seems that the Aces have magically travelled across.

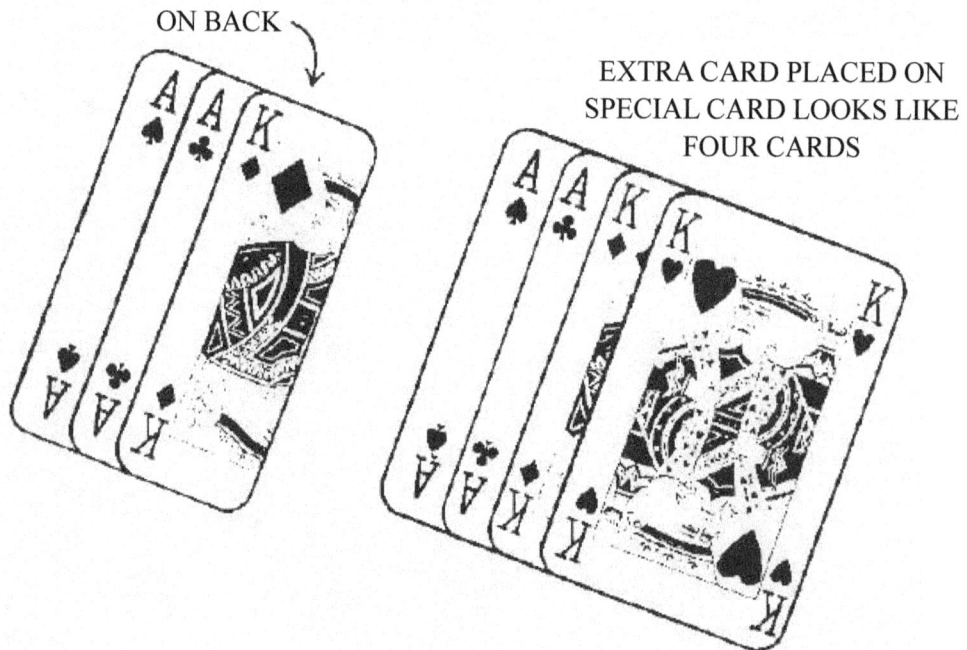

ON BACK

EXTRA CARD PLACED ON SPECIAL CARD LOOKS LIKE FOUR CARDS

A Spelling Bee

Effect: Ten cards are counted in a magical manner.

Secret: The cards you use have been secretly arranged in the order, as shown. The actual suits do not matter. Run the cards, one at a time, from hand to hand. This reverses the order so the Nine is now on the top. Place the 10 cards on the pack. To

on the E and it is an Ace. Drop this card on the table. Now spell T W O – again moving one card from the top to the bottom of the packet for each letter. Turn over the "O" card, a Two, and drop it on the table. Continue spelling in this way until you get to 10. For the 10th card, you have only one card left but pretend to spell the letters as before to amuse your audience.

14

A Great Escape

Effect: You escape from a rope.

Secret: Ask someone to tie your wrists together with a large scarf. A long length of rope is now placed between your wrists and someone holds the ends. Another scarf is thrown over your bound wrists. As soon as your hands are out of sight, move your hands back and forth to cause the rope to form a loop between your wrists. Keep working the rope until you can get one hand into the loop. When you have done this, ask the person who is holding the rope to pull – the rope will come free and yet your wrists are still securely tied!

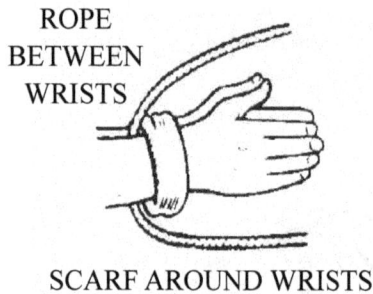

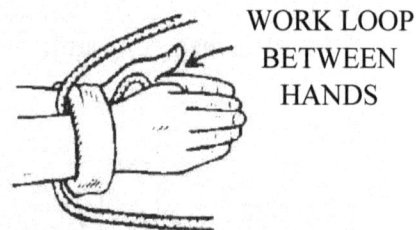

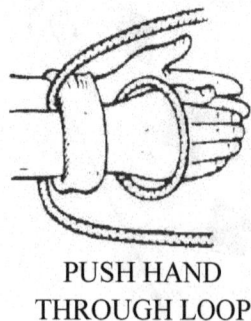

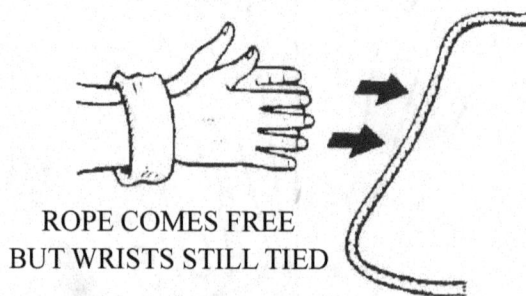

ROPE
BETWEEN
WRISTS

WORK LOOP
BETWEEN
HANDS

SCARF AROUND WRISTS

PUSH HAND
THROUGH LOOP

ROPE COMES FREE
BUT WRISTS STILL TIED

A Disappearing Pencil

Effect: A pencil disappears.

Secret: Show a pencil or pen and drape a large handkerchief over it. As soon as the

allow the pencil to drop down your sleeve. Take the handkerchief with the other hand as if holding the pencil through the material. Throw the handkerchief into the air and the pencil has vanished!

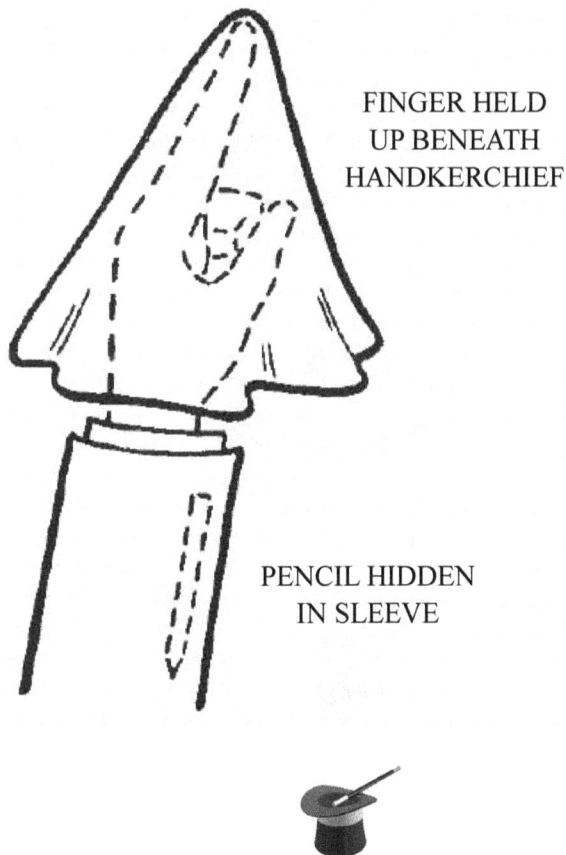

FINGER HELD
UP BENEATH
HANDKERCHIEF

PENCIL HIDDEN
IN SLEEVE

Baffling Bands

Effect: Three loops of paper are cut with surprising results.

Secret:

strip but give one end a half turn before gluing the ends. In making the third loop,

with a pair of scissors. This will make two loops as one would expect. When the second loop is cut in the same way, it forms one extra large loop and the third one produces two loops linked together.

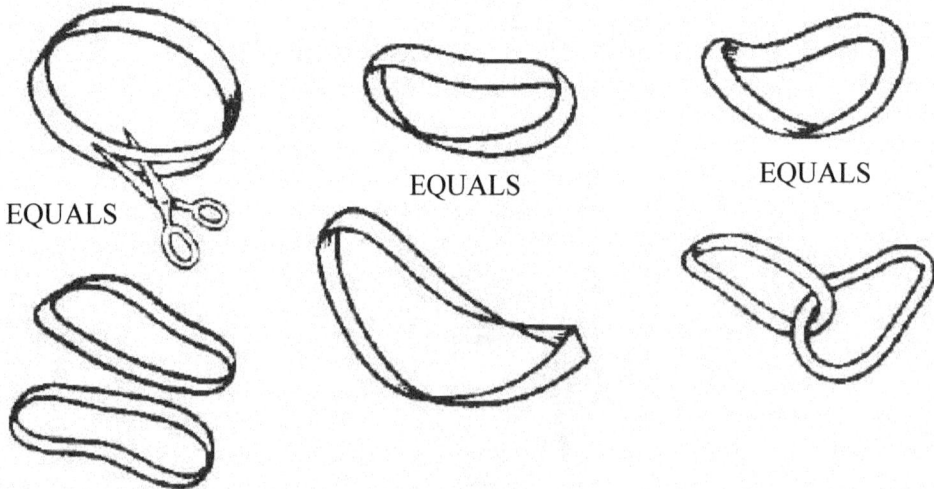

EQUALS

EQUALS

EQUALS

Jean Eugène Robert-Houdin

Master of magic Robert-Houdin was born as Jean Eugène Robert in Blois, France, on December 6, 1805—a day after his autobiography said he was. His father, Prosper Robert, was one of the best watchmakers in Blois. A skilful artisan and hard worker, Prosper Robert's main ambition was to provide for his family, but he also wanted his children to climb the social ladder. Jean Eugene's mother, the former Marie-Catherine Guillon, died when Jean was just a young child. At the age of eleven, Prosper sent his son Jean to school thirty-five miles up the Loire to the University of Orléans. At 18, Jean graduated and returned to Blois. His father wanted him to be a lawyer, but Jean wanted to follow into his father's footsteps as a watchmaker.

In the mid-1820s, young Jean saved up to buy a copy of a two-volume set of books on clock making. Written by Ferdinand Berthoud. The bookseller had put the books off to the side for Jean. He reached up to the shelf and grabbed the books. He wrapped the two volumes and handed them to the young aspiring clockmaker.

When Jean got home and opened the wrapping, instead of the Berthoud books, what appeared before his eyes was a two-volume set on magic called Scientific Amusements. Instead of returning the books, his curiosity got the best of him. From those crude volumes, he learned the rudiments of magic. He practiced at all hours of the day.

He was upset that the books he got only revealed how the secrets were done, but did not show how to do them. So Jean began taking lessions from a local amateure magician who taught him how to juggle to coordinate his eye and hand.

He also taught him rudiments of the cups and balls.

Magic was his pastime, but meanwhile, his studies in horology continued. When he felt he was ready, he moved to Tours and set up a watchmaking business, doing conjuring on the side.

Much of what we know about Robert-Houdin comes from his memoirs.

His early performance came from joining an amateur acting troupe. Later, he performed at social parties as a professional magician. It was during this period while at a party that he met the daughter of a Parisian watchmaker, Monsieur Jacques François Houdin, who had also come from Jean Robert's native Blois. The daughter's name was Josèphe Cecile Houdin, and Jean fell in love with Cecile at their first meeting. On July 8, 1830, they were married. He hyphenated his own name to hers and became Robert-Houdin.

He moved to Paris and worked in his father-in-law's wholesale shop. While M. Houdin worked in the main shop, Jean was to tinker with mechanical toys and automatic figures.

With his work in the shop, Jean was still practising magic. Quite by accident, Robert-Houdin walked into a shop on the Rue Richelieu and discovered it sold magic. There, he met fellow magicians, both amateur and professional, where he engaged in talk about conjuring, and he met an aristocrat by the name of Jules de Rovère, who coined the term "prestidigitation" to describe a major misdirection technique magicians used.

At Papa Roujol's, Robert-Houdin learned the details to many of the mechanical tricks of the time as well as how to improve them. From there, he built his own mechanical figures.

On October 19, 1843, Monsieur Robert-Houdin's beloved wife died, having been ill for months; she died at the age of thirty-two. At her death, she left him with three young children to take care of; to take up the burden, he remarried in August of that year to François Marguerite Olympe Braconnier, a woman ten years younger than himself. The new Madame Robert-Houdin soon took over the household.

Robert-Houdin loved to watch the big magic shows that came to Paris. He dreamed about some day opening his own theatre. In the meantime, he was hired by a friend of his by the name of Count de l'Escalopier to perform at private parties.

Now that he had free time, he began constructing equipment for his own use instead of selling it to others. The income from the shop and his new inventions gave him enough money to experiment on new tricks using glass apparatus that would be (or at least appear) free of trickery. He envisioned a stage that would be as elegant as the drawing rooms in which he was hired to perform. He also thought that a magician should be dressed as such by wearing traditional evening clothes.

On July 3, 1845, Robert-Houdin premiered his 200-seat theatre in what he called "Soirées Fantastiques". No critics covered Robert-Houdin's debut, and in his memoirs, Robert-Houdin said that the show had been a disaster. He suffered from stage fright that caused him to talk too fast and in a monotone.

After the first show, he was about to have a nervous breakdown. He closed the theatre and had every intention to close it for good, until a friend agreed that the venture was a silly idea. Instead of admitting defeat, Robert-Houdin, irked at the friend's affrontery, used this insult to regain his courage, and persevered in giving the show a long run at his little theatre. Although at first, the forty-year-old magician was unpolished, he soon gained the confidence required for the stage.

With each performance, Robert-Houdin got better, and he began to receive critical acclaim.

The following year, he added a new trick to his programme that became especially popular. Seats at the Palais Royal were at a premium. This new marvel was called Second Sight. Second Sight drew the audiences into the little theatre. Once there, they saw the other creations Robert-Houdin had to offer.

The Arabs of Algeria were said to be excited to rebel against French colonialists by miracles performed by their religious leaders. In 1856, Napoleon III's Second French Empire sent Robert-Houdin there, hoping that he might perform tricks that were far more impressive, thereby dissolving the excitement of the rebels. Robert-Houdin's tricks, it is said, succeeded in breaking up the influence of the mullahs. Moreover, the Arabs became afraid of Robert-Houdin. In one trick, he allowed an Arab to shoot at him with a marked bullet, but instead of killing him, the bullet was found between his teeth. After that, they believed he could do anything. Robert-Houdin was not the first illusionist to perform the bullet catch, and many since him have adapted their own version of the effect.

He used another famous trick to prove that French magic was stronger than local shamanism techniques: he presented an empty box with an iron bottom that anyone could lift. By turning on an electro-magnet hidden under the floor, he made it immovable, "proving" that through will power, he could make it impossible to lift the trunk for the strongest Algerian warriors. He found the trick was more impressive not when he claimed that he could make the trunk heavy, but when he claimed he could make the strong men too weak to lift a trunk that even a small child could lift.

Robert-Houdin is often credited as being "the father of modern magic". Before him, magicians performed in marketplaces and fairs, but Robert-Houdin performed magic in theatres and private parties. He also chose to wear formal clothes, like those of his audiences.

It`s Matchic!

Effect: You reveal how many matches were moved while your back was turned.

Secret: Secretly put a small pencil dot on a match, just below the head. Do this on all four sides of the match and then put if back in the box. When you show the trick, remove ten matches from the box and put them down in a row. The marked match must be placed at the left end of the row. Turn your back and ask someone to move any number of matches, one at a time, from the left end of the row to the right. When you turn around, you have only to spot the position of the marked match and count from that to the right end of the row to know how many matchsticks have been moved.

A MARKED MATCH AT THE LEFT END OF ROW

PENCIL
DOT

POSITION OF THE MARKED
MATCH SHOWS THAT IT HAS
BEEN MOVED

A Jumping Elastic

Effect:

Secret:

into the band as shown in the second drawing. Open your hand and the elastic band

your left hand towards your audience at all times when performing this trick and it

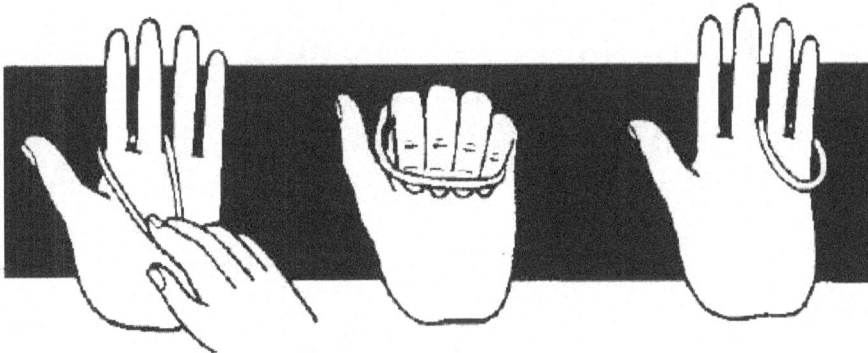

Walking Through A Postcard

Effect: A postcard is cut so that it will go over your body.

Secret: To show that you can walk through a postcard, make the following moves. Fold the postcard in half lengthways. Make as many cuts as possible from the edge of the card to the centre and from the centre towards the edges, as shown. The more cuts you make, the easier the trick is to do. Unfold the card and cut along the centre, from A to B. You can now open out the card into a large loop that will easily go over your body. Wow! This is interesting. Isn't it?

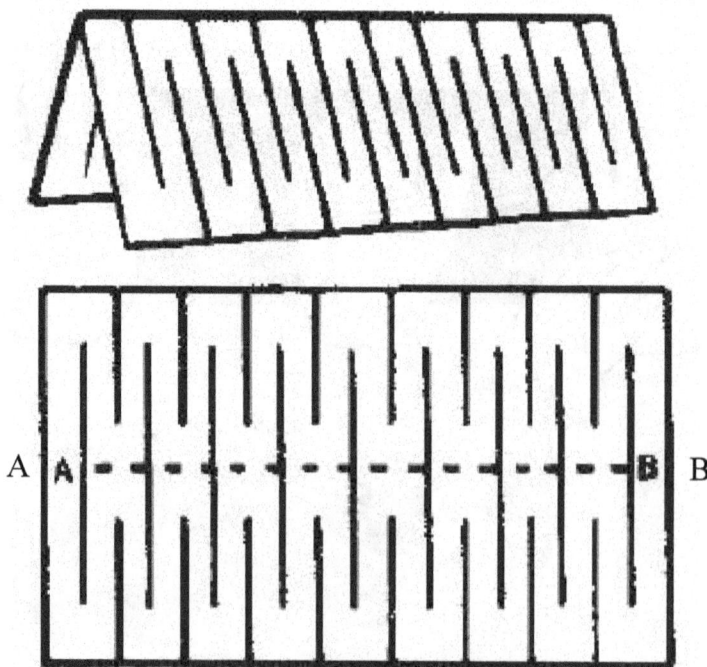

A A - - - - - - - - - B B

20

A Sympathetic Diamond

Effects: A large Ace of Diamonds in cards changes colour when coloured handkerchiefs are passed over it.

Secrets: The principal Diamond on the card is actually cut out. It gets its colour from a piece of card held behind the Ace. Each time you want the colour of the diamond to change, the secret card is moved to a new position. To get a multi-coloured diamond, position the card centrally so that all the four corners on the card is visible through the diamond-shaped hole. The handkerchiefs hide the movement of the secret card, look colourful and provide a reason for the colour changes.

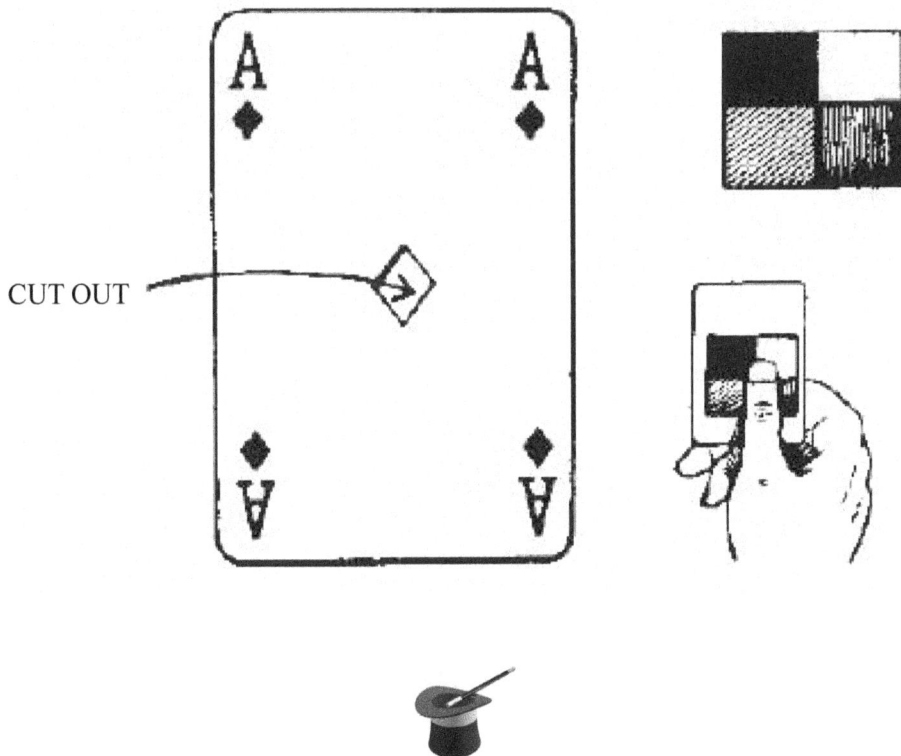

CUT OUT

21

A Wallet Of Wonders

Effects: A one Rupee coin changes to two 50-Paisa coins.

Secret: To make the special wallet, you require two sheets of paper, folded, as shown. Open out the folds and glue the shaded areas together. Put two 50 Paisa coins in the centre of the top sheet and fold it around them. Fold up the bottom sheet and turn the whole lot over. Place a borrowed one Rupee coin in the centre of the empty sheet and fold the paper around it. As you do these secretly turn the whole packet over. Open the paper to reveal the two coins. This trick requires quick and intelligent moves with your hand.

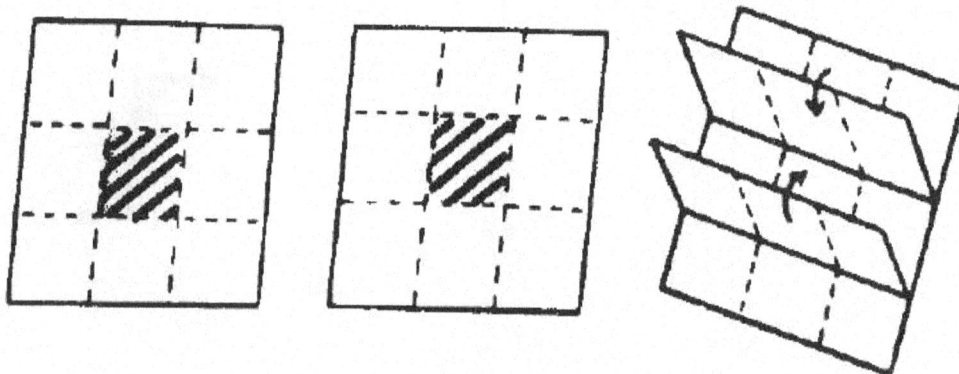

Prof. K. Bhagyanath

His show was known as 'The biggest show of magical entertainment in South India'. Prof. K. Bhagyanath was born in 1916. K. Bhagyanath, is the sole architect of FANTASIA - a super show of magic and illusions. Fantasia Centre is dedicated to the memory of this great man who left us on January 15, 1999.

Magic was his passion from the tender age of 9. He gave his first performance as an amateur magician at the age of 12. Though magic was his real passion, he finished his masters degree in literature and became professor of English. During early 1960, his passion overtook his profession and resigned his job to become a full-fledged magician.

In 1963, he shifted bag and baggage to Madras (Chennai) giving up the security of a permanent job a to launch his FANTASIA, a huge show of magic and illusion in 1965.

His most remarkable attribute was his transparent sincerity and his response to any situation was governed by textbook idealism. During his young ages, he became influenced by Mahatma Gandhi and took up the Indian freedom struggle to his heart. He was a prominent student leader throughout the independence struggle of India.

Tall, handsome and with a voice that needed no megaphone, his shows ran to overflowing houses. Often quoted as "The most educated Magician in the World", he lived upto it. His enthusiasm for research and experimentation on magic and allied arts was phenomenal and his collection of books on this subjects is a treasure house of information.

He believed that magic is one of the finest arts. He used to say, "All arts create illusions." Magic creates illusions but with a difference. While all other arts try to create the illusion of reality, magic alone creates the illusion of unreality of supernatural and of unseen powers at work.

Link Up

Effect: Two paper clips linked together without and physical contact.

Secret: You need two paper clips and a sheet of paper. Fold over one-third of the paper and place a clip over the top of the doubles thickness. Fold the other end of

two thicknesses of paper just formed. The position should now be as shown. Pull sharply on each end of the paper and the clips will leap off into the air and link themselves together without the help of your hands!

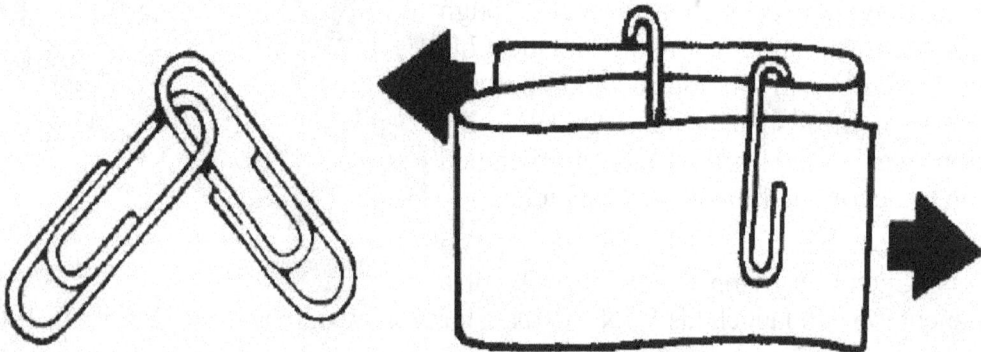

Mind Reading Trick

Effects:

return, you read your assistant's mind and reveal the number.

Secret: When you return to the room after a number has been chosen, you place your hands on either side of your assistant's head. Your friend now tightens and loosens his jaw for the appropriate number of times to signal each number in turn, pausing

movement, and so becomes easy for him to transmit numbers to you without saying

A Coin in Bread Roll

Effect: A bread roll is broken open and a coin is found inside.

Secret:

this, so keep it hidden. When you are at the dinner table take a bread roll and place it directly on top of the concealed coin. Bend the roll downwards between your hands. This causes an opening in the bottom of the roll. Force the coin into this opening. Now break the roll from the top. As the roll breaks, the coin appears from the middle of the roll.

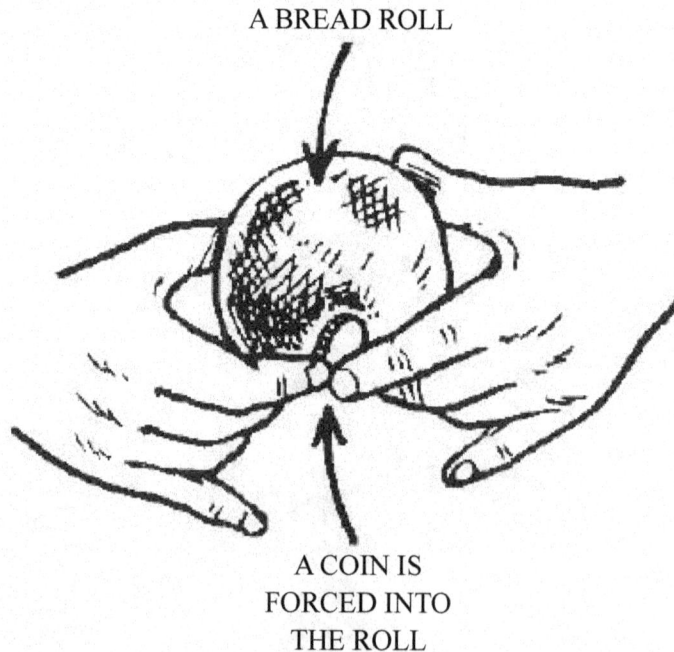

A BREAD ROLL

A COIN IS
FORCED INTO
THE ROLL

Flag of the Realm

Effect:

Secret: Fold a Union Jack and place it on a sheet of newspaper. Put another sheet

them. The prepared newspapers are on your table at the beginning of the trick. Show some red, white and blue ribbons. Wrap them up in the newspaper. Tear into the

Note: The trick can be converted into the Indian Flag also; all you need is Orange, White and Green ribbons and of course, the Indian Flag.

FOLDED UP
FALG

26

Turn over Card

Effect: A chosen card reverses itself in the pack.

Secret: Spread the cards face down and ask a spectator to take any card. While he is showing the card to the rest of the audience, secretly reverse the bottom card and then turn the whole pack over in your hands. Walk towards the spectator and ask him to replace his card in the pack, face down. As you walk back to your table, turn the reversed card face up and turn over the whole pack again. All the cards are now face down with the exception of one card – the one that the spectator chose.

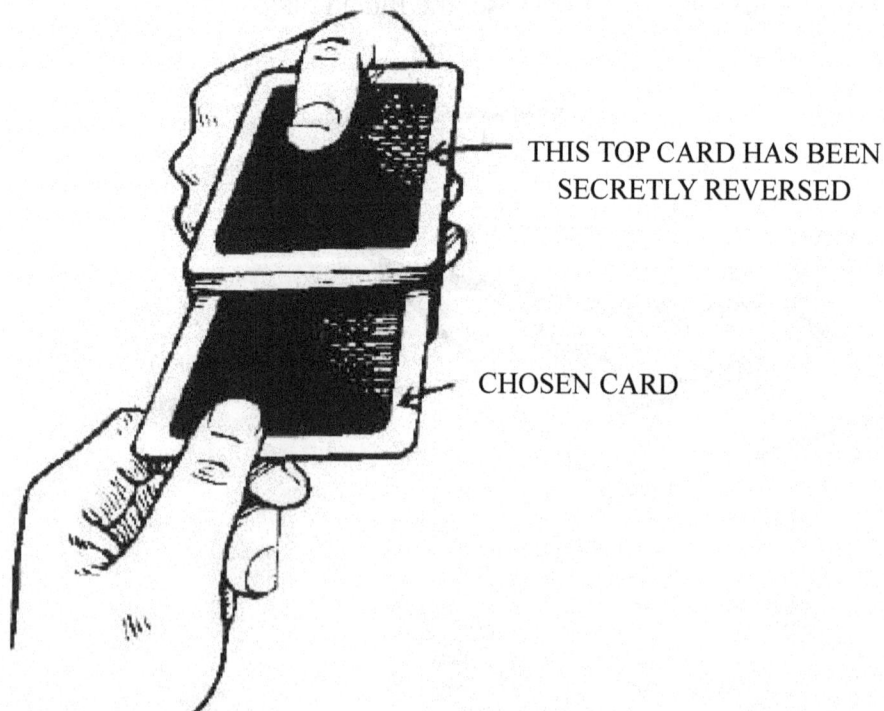

THIS TOP CARD HAS BEEN
SECRETLY REVERSED

CHOSEN CARD

David Blaine

David Blaine was born on April 4, 1973 in Brooklyn, New York. He is of Puerto Rican and Jewish decent and grew up in New Jersey, attending Passaic Valley High School in Little Falls. Blaine first became interested in magic at age 4 after he saw a street magician perform card tricks. David's mother Patrice White strongly encouraged his new interest.

In his teens, Blaine caught the acting bug and frequently commuted from New Jersey to New York City to attend acting classes and auditions, managing to land some work in commercials and on the soaps.

At age 20, Blaine learned that his mother had been stricken with ovarian cancer. She passed away in 1994, but Blaine is frequently quoted saying that she remains an important influence on his life.

On his own, Blaine eventually settled in the Hell's Kitchen neighbourhood of Manhattan and continued to pursue his dream of becoming a famous magician.

The ambitious Blaine made sure to attend the celebrity functions and mingle with the rich and famous. He performed tricks for the likes of Leonardo DiCaprio, Robert DeNiro, Al Pacino and David Geffen. Word about his prowess as an illusionist soon began to spread.

Early on, David Blaine made his name by performing magic tricks on the street and filming his volunteers' reactions. The most dazzling of his street tricks was his levitation illusion, in which he appeared to magically raise himself up to two feet above the ground.

Hoping to drum up interest in his act, David sent a tape of his street performances to ABC. The response was encouraging to say the least. ABC gave Blaine a million dollar contract to produce David Blaine: Street Magic in 1997, which was followed by David Blaine: Magic Man in 1999.

The Performances/Spectacles

Buried Alive

On April 5, 1999, Blaine spent one week buried inside a glass coffin at Donald Trump's Trump Place Development on the West Side of Manhattan. Except for an area at the top that was open to the viewing public, the coffin was completely covered with earth.

Frozen in Time

On November 27, 2000, Blaine was encased in ice for almost 62 hours in New York's Times Square. A network of tubes provided him with air, water and the ability to relieve himself.

Vertigo

On May 22, 2002, Blaine spent over 34 hours on a pillar, 90 feet high and 22 inches wide in New York's Bryant Park. David sustained a concussion during the stunt when he dove into a large pile of cardboard boxes at the foot of the pillar.

Above the Below

On September 5, 2003, Blaine spent 44 days sealed inside a transparent glass box suspended 30 feet above the ground in London's Potter Fields Park on the bank of the River Thames. The glass box was approximately 7 feet x 7 feet x 3 feet. During the 44 days, Blaine claims to have fasted and lost 54 pounds of body weight.

Drowned Alive

On May 1, 2006, Blaine spent one week immersed in an 8-foot transparent sphere filled with water at Lincoln Centre in New York City. The sphere was filled with 2000 gallons of water at a temperature of 96 degrees. Upon completing the week-long endurance test, Blaine removed his breathing apparatus in an attempt to break a world record of almost nine minutes without breathing. Unfortunately, he only reached 7 minutes 8 seconds before needing to come out. To prepare for the stunt, Blaine claims to have dropped his weight by 50 pounds so that his body would require less oxygen.

Gyroscope

On Tuesday November 20, 2006, Blaine was suspended 40 feet in the air above Times Square while spinning 8 times per minute. The stunt lasted for 52 hours as Blaine escaped the spinning gyroscope and managed to free himself on Thursday, November 22, at about 2.15 pm. He had no protection form chilly, rainy weather other than his clothing.

Mystifying Pencil

Effect: A pencil adheres to one hand and then to the other.

Secret: The secret is a small pin pushed into the pencil. Hold the pin between the

hand to the pencil. Turn to the left and revolve the pencil between your hands. This

this movement, you can move the pencil back to the right hand. This trick needs caution as you are using a pin and a pointed pencil. At the same time, it needs real quick movements.

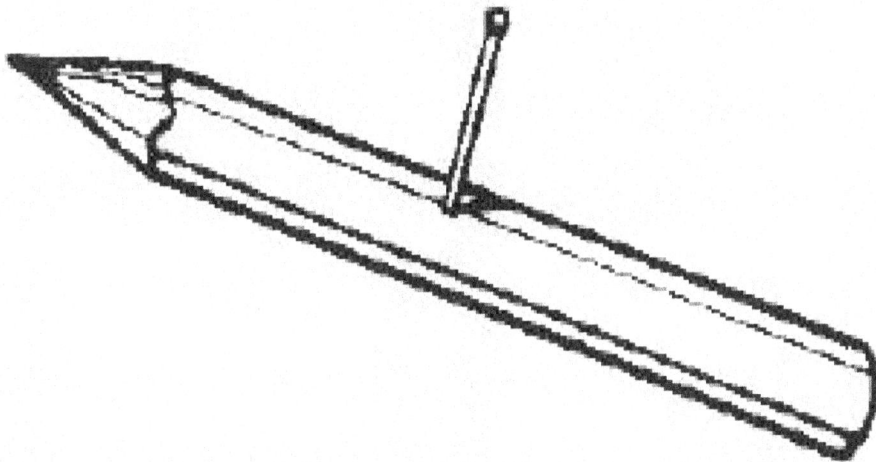

A Royal Flush

Effect: Five cards change into a Royal Flush.

Secret: Take the Ten, Jack, Queen and King of one set of cards and cut them into half diagonally. One half is glued on to one of four other cards of any value and suit.

the cards so that the backs are towards the audience. Take off the Ace and show it, as you secretly turn the other cards end for end. Replace the Ace and then let everyone see the faces. They have changed into a *Royal Flush*.

Money Making Matchbox

Effect: Two coins are placed in a matchbox and the box is closed. When the box is opened again, the number of coins in it has increased to three.

Secret: Open the matchbox and hide a coin between the top of the drawer and the cover of the box, as shown. Place two coins in the box and shut the drawer. This forces the hidden coin to drop into the drawer. Make a few mystical passes over the

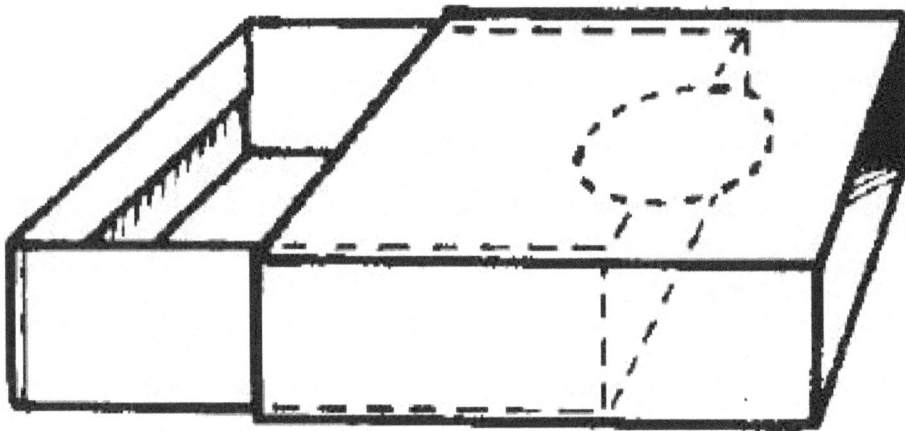

it?

An Impossible Escape

Effect: A handkerchief escapes from a sealed tumbler.

Secret:
to one corner which hangs over the outside of the glass. A second handkerchief is

with an elastic band. Reach under the handkerchief covering the glass and pull the secret thread, pulling the coloured handkerchief from the glass into your hand. Pull it into full view and then remove the elastic band covering the handkerchief to prove and that the coloured handkerchief really did escape. This trick also needs quick and careful hand movements.

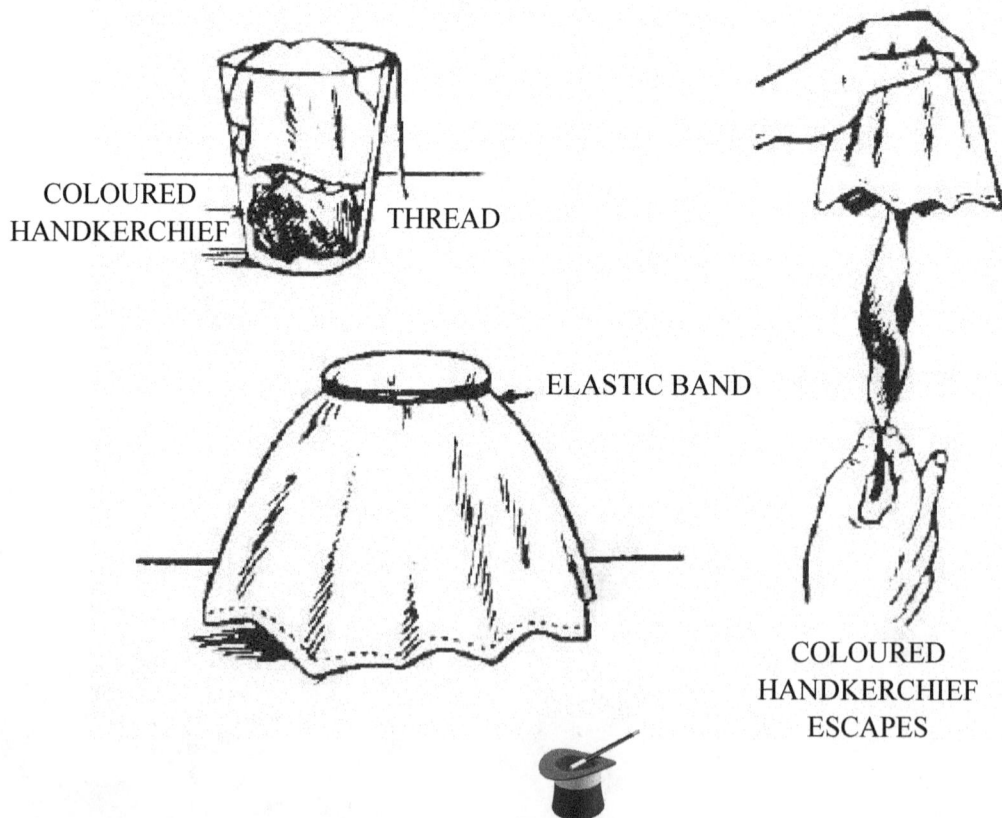

COLOURED
HANDKERCHIEF   THREAD

ELASTIC BAND

COLOURED
HANDKERCHIEF
ESCAPES

The Magical Seven

Effect: You predict which pile of cards a spectator will choose.

Secret:

is then folded and given to a spectator for safekeeping. You now take several cards from a pack and arrange them in three face-down piles. The three piles are formed

any of the three piles. Whichever pile is chosen, the prediction is always correct.

7 CARDS

THE FOUR 7'S

ADDS UP TO 7

P.C. Sorcar

Pratul Chandra Sorcar (P C Sorcar Sr.): Pratul Chandra Sorcar (P C Sorcar Sr.) is known as the Father of Modern Indian Magic. Magic being in his blood as he profoundly asserted. "When asleep I breathe Magic; when awake I work Magic," Sorcar said in full spirit of ease and delight.

The very name of P C Sorcar conjures up the vision of outstanding feats of Indian magic: the Rope Trick, the Flying Carpet, the X-Ray Eyes. He cast his spell over the most unbelieving audiences and showed before the naked eyes what people thought was impossible.

Besides personal achievements in honours (Maths), fame and glory Sorcar's greatest contribution to the world of magic was the installation of Indian magic - his beloved Ind-dra-jal - to the pedestal of pristine glory with greater halo round its crown. Under his light and lead, it became an Art of International attraction. Sorcar's triumphal success was a magic by itself, but not achieved in one day. It was a whole life's dedication that made him the world's greatest magician. He went around the globe several times performing his magic lifting a dying art of India and bringing it to limelight. His items were not mere tricks; each was a combination of modern science mixed with tastefully designed art. Micky Hades, the editor of Hade-E-Gram monthly magazine of Calgary, Canada, wrote about him in his article titled "Sorcar's Artistic Triumph": He was the founder of All India Magic Circle where thousands of members from around the world learned magic under his guidance.

Born in a family of magicians of seven generations, Sorcar started off as a stage name for Protul Chandra Sorcar, a name destined to attain ranks of immortals later. A citizen of India, he was born on February 23, 1913 in the small town of Tangail in Mymensing which now lies in Bangladesh. His father's name was Bhagawan

Chandra Sorcar and mother, Kusum Kamini. He had one sibling, a brother, Atul Chandra, ten years younger than him. Sorcar was a brilliant student at school. He graduated from Tangail Shibnath High School in 1929 with first class. In 1931 he earned his I.A. (Intermediate in Arts) degree from the Karotia College (first class) and then joined Ananda Mohan College for B.A. (Bachelor of Arts) with honours in Mathematics. In 1938 he married Basanti Devi, the daughter of Dr. Pramatha Nath Majumder, a renowned medical doctor of Mymensing, Bangladesh. Basanti Devi remained the main source of inspiration for all his achievements throughout Sorcar's life.

From the very childhood Sorcar found magic to be the passion of his life, which he took up as a full time profession after he sat for his B.A. degree tests in 1933. His singular devotion soon brought its own honest reward. His unique feats of the newly cultivated art had soon won robust acclamation from the press and the public alike. He was hailed as one giving to the Art of Indian Magic a new cultural background which readily found a strong international appeal.

Sorcar's interest in writing books on magic flourished simultaneously with his magic shows. Over his life time, he has been a regular contributor of magic-articles to numerous magazines and journals thoughout the country and was the author of over 22 books on magic, starting with "Hypnotism", which he wrote while still a college student. The success of the first book led to the other books such as "Mesmerism", "Chheleder Magic", "Magiker Kaushal", "Sahaj Magic", "Magic For You", "More Magic For You", "Hindu Magic", "100 Magic You Can Do", "Indrajal", "Deshe Deshe", "Sorcar on Magic", "History of Magic" etc. in Hindi, Bengali and English languages.

Sorcar died of a heart attack at the young age of 58 in Ashaikawa, Hokkaido, Japan, on January 6, 1971, where he was performing his Ind-dra-jal. The world mourned the great king of magic. Messages of condolence came from friends, families, magicians from all over the world, and from Government authorities and leaders of many countries including Japan, U.K., Australia, New Zealand, USA, and Soviet Union. India's Prime Minister Indira Gandhi mourned the loss of the great son of India stating "...with the death of Mr. Sorcar has ended the glorious chapter of Indian magic".

He was awarded the Padma Shri by the President of India in 1964 and The Sphinx known as Oscar of magic twice in 1946 and in 1954.

A Magnificent Memory

Effect: You appear to memorise the order of a pack of cards.

Secret: Cut a small hole in the bottom right corner of your card case. Ask someone

then place them in the case. Hold the case in front of you and you will see the index

same with a few more until you have proved that you really did memorise the order
of the cards.

GUT OUT
WINDOW

A Calculated Discovery

Effect: You remove the cards from your pocket to the value of a chosen card.

Secret: Put the Ace of Clubs, Two of Hearts, Four of Spades and Eight of Diamonds

name a card. The four cards you hid can represent any card. For the Six of Clubs, you take out the Four and the Two to make six and the Ace to show the Clubs. For the Queen of Diamonds, take the Eight and the Four to make 12 and point to the

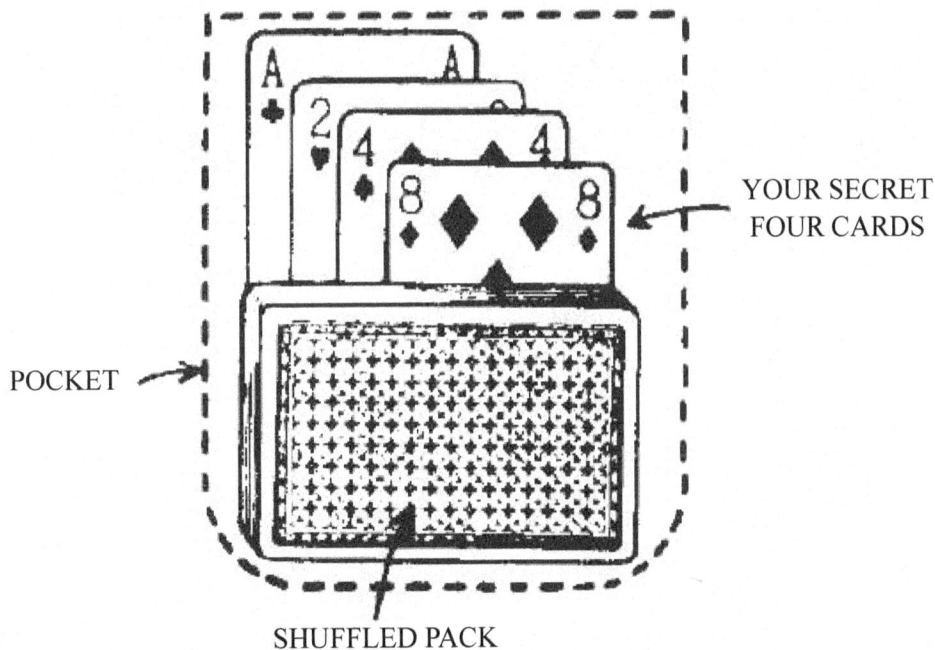

YOUR SECRET
FOUR CARDS

POCKET

SHUFFLED PACK

34

A Bangle of Deception

Effect: A bangle passes on to a length of rope tied securely to your wrists.

Secret: You need two identical bangles and a long piece of soft rope. Secretly put one of the bangles on your arm, hidden under your coat sleeves. Have two spectators tie each end of the rope around your wrists. Show the bangles to ensure that it is a solid. Turn your back, and quickly pull the secret bangle down your sleeve on to the rope and hide the bangle just shown in an inside pocket. Turn back to face the audience and show the bangle tied on to the rope. This needs real quick and clever movements of your hands.

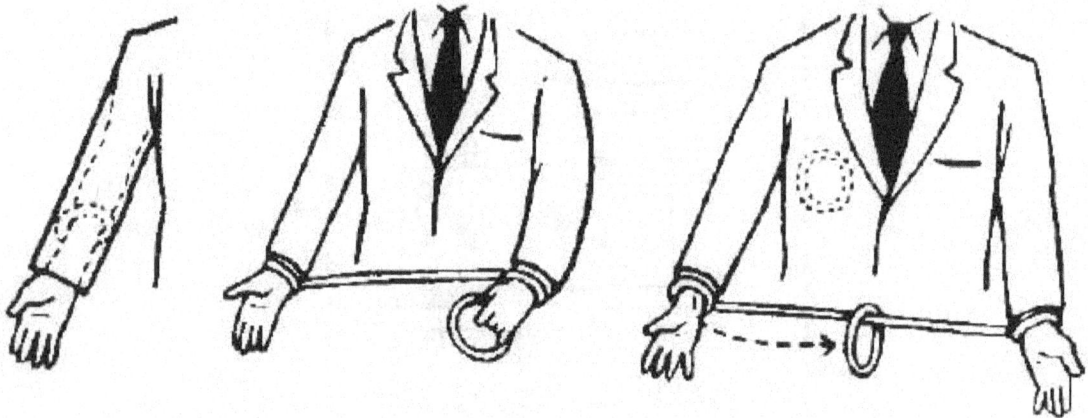

An Obeying Ball

Effect: A ball rolls across the table until you tell it to stop.

Secret: Under the table cloth is a small ring attached to a length of strong thread. The thread runs across the table to where your secret assistant is sitting. A small ball is placed on the table. It must go into the concealed ring. On your word of command, your assistant pulls on the thread and the ball moves. You then pick up the ball and hand it to a spectator. As you are doing this, your assistant pulls the ring from beneath the cloth and hides it. This is how a ball follows your command!

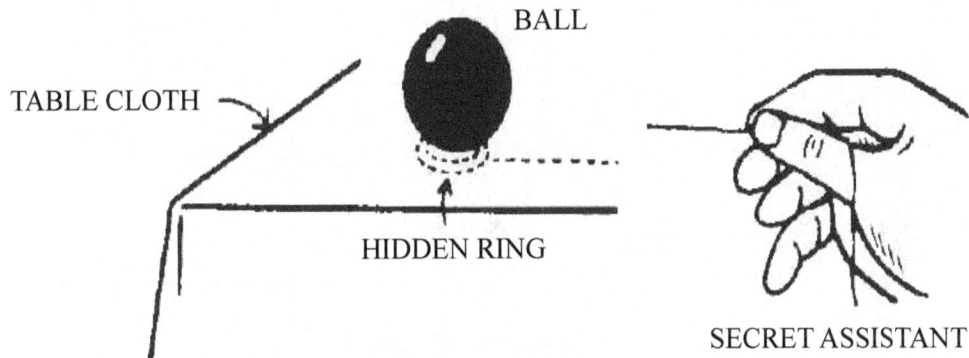

BALL

TABLE CLOTH

HIDDEN RING

SECRET ASSISTANT

Fruit Transformation

Effect: An orange transforms itself into an apple. This is really interesting!

Secret: Carefully cut the peel from an orange, as shown in the illustration and scoop out all the fruit. When the peel has dried, put an apple inside it. Hold the peel closed and it appears you are holding an ordinary orange. Cover the orange with a handkerchief. When you remove the handkerchief, take the orange peel off the

as you display the apple. Wow! You have changed an orange into an apple with your magic.

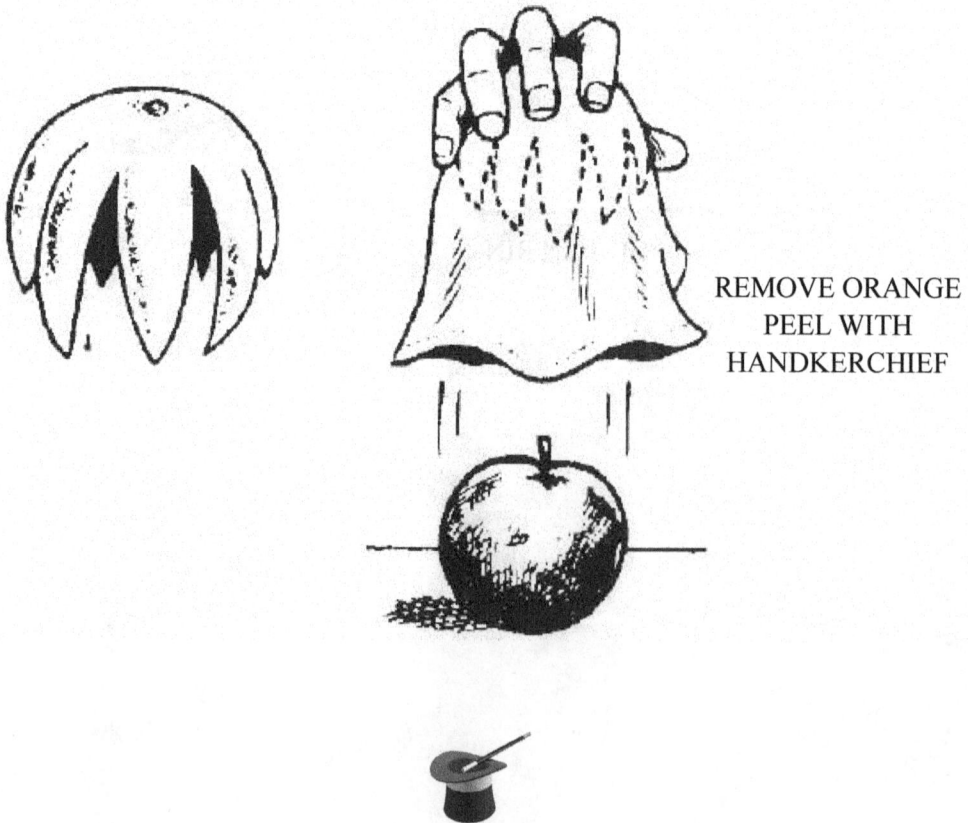

REMOVE ORANGE
PEEL WITH
HANDKERCHIEF

Kuda Bux

Kuda Bux (1906 - February 5, 1981) was an Indian mystic and magician. He was born in Aknmur Kashmir, Pakistan as *Khudah Bukhsh*. One of his most famous tricks was one in which he would cover his eyes with soft dough balls, blindfold himself, swath his entire head in strips of cloth, and yet be able to see. He was also a fire walker. In his later life, he lost his eyesight to glaucoma.

He was one of the few who had the ability to use the felicity of the brain which most of us never ever think of. Kuda Bux was the subject of a 1950 film titled *Kuda Bux, Hindu Mystic*, and his apparent ability to see while blindfolded with dough balls strongly influenced British author Roald Dahl in the short story of Henry Sugar, who was taught to develop the same powers.

Most astoundingly, observers noted that the unblindfolded Bux required reading glasses to read fine print. Whilst blindfolded Kuda Bux would read the dates on coins which are held on a spectator's hand, read the fine print of a magazine, thread a needle, duplicate words he had never seen written, shoot a bull's – eye with a pellet gun, and many other mysteries.

According to Robert Ripley, Kuda Bux performed an astonishing feat in Radio City, Manhattan, on August 2, 1938. According to this account, a hole 3 – feet deep was dug in the Radio City parking lot and logs and bags of charcoal were set on fire in it. Kuda Bux, so the story goes, walked back and forth through the pit—twice. Ripley said, "Kuda Bux's feet were not even warm." There is newsreel footage of this event in the TV biography (distributed on VHS) Robert Ripley: Believe It or Not (TBS 1993).

He performed another astonishing feat in 1935 when he demonstrated fire walking (another one of Kuda Bux's amazing talents) in front of an audience of scientists from the University of London Council for Psychical Research and news reporters. He walked across a twelve foot pit of burning hot coals unscathed. Bux's feet were checked before and after the fire walking demonstration to verify that no protective chemicals, topical creams or herbs were used. It was a very windy day and the surface temperature of the fire was read at 806 degrees Fahrenheit. The body of the fire was measured at 2,552 degrees Fahrenheit which is hot enough to literally melt steel. After Kuda Bux's firewalking through the coals, a cameraman who had messed up some photographs of the event, asked for a retake. Kuda Bux obliged by repeating the firewalk once more. Again, his feet were checked before and after the fire walking demonstration.

A Chosen Card Trick

Effect: You instantly identify a selected card.

Secret: Place all the black cards on top of red cards and you are ready. Spread the cards face down and ask someone to take any card. Make sure he takes one from the

Look through the cards, and somewhere in the red cards will be one black card. That is the *chosen card*. Remove it and show it to everyone. Great, isn't it?

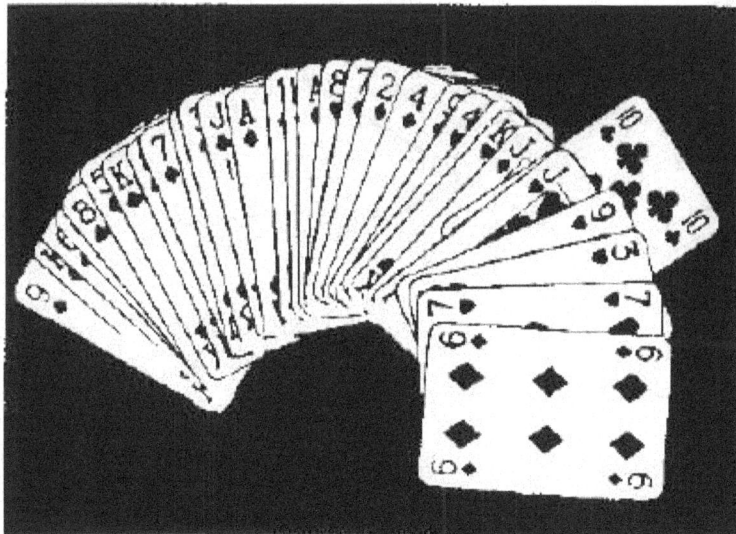

THE ONLY
BLACK CARD
IN RED
SECTION

BLACK SECTION   RED SECTION

Going Up!

Effect: Two tumblers are suspended from a newspaper balanced on a magic wand.

Secret: Bend a paper clip around your magic wand to form a U shape. Tape the

The secret wire should be on the side of the newspaper nearest to you. Place the newspaper across two tumblers, making sure that each prong of the wire goes into the tumblers. Place your wand between the tumblers and lift it towards the newspapers. The newspaper will be lifted – but so will the tumblers. This is how you can lift the newspaper along with the tumblers with your magic wand!

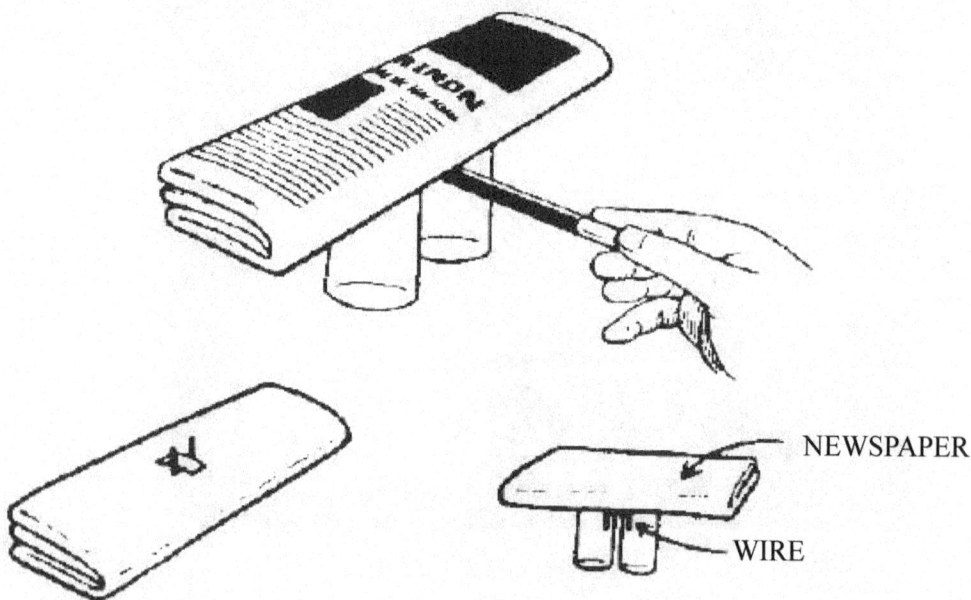

NEWSPAPER

WIRE

Name Game

Effect: You reveal a thought-of name. This magic has to be done very cleverly.

Secret: Cut a slot in the top of a box with a removable lid and tape a small box on the inside. Write the same name on 10 slips of paper. Put them in the box and replace the lid. Hand out 10 blank slips and ask the spectators to write a name on their slip. The slips are put in to the box through the slit. They actually go into the hidden box. Remove the lid for just one slip to be chosen. As all the slips are the same, it is easy to reveal the name chosen.

Note: Be tactful that no one comes closes or handles the box, except you.

The Ninth Card

Effect: You reveal the identity of a selected card.

Secret: This is how to make someone take the card you want. Secretly place the card you decide to use the ninth from the top of the pack. Ask someone to call a number between 10 and 20. Count that number of cards, one at a time, from the pack. Add the digit of the chosen number. Deal that number of cards off the cards just dealt. Ask someone to look at the next card. It will be the card you placed in the ninth position. You can reveal it in any way you wish.

THE FORCED CARD IS
PLACED NINTH FROM
THE TOP OF THE PACK

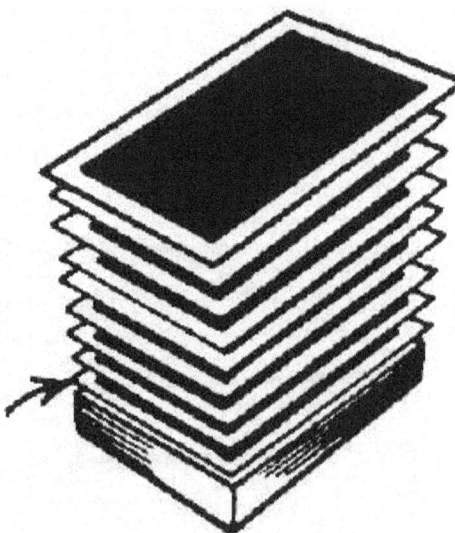

An Unbreakable String

Effect: A string is threaded through a drinking straw. The straw is cut in two but the string remains unharmed.

Secret: Secretly cut a short slit in the straw. Thread the string through the straw. Bend the straw in half and pull both the ends of the string. This forces the string through the slit in the straw. Your hand hides this from the audience. Cut through the straw, as shown by the dotted line in the drawing. Now pull the string from your hand. The straw is in two pieces but the string is still whole. Amazing, isn't it?

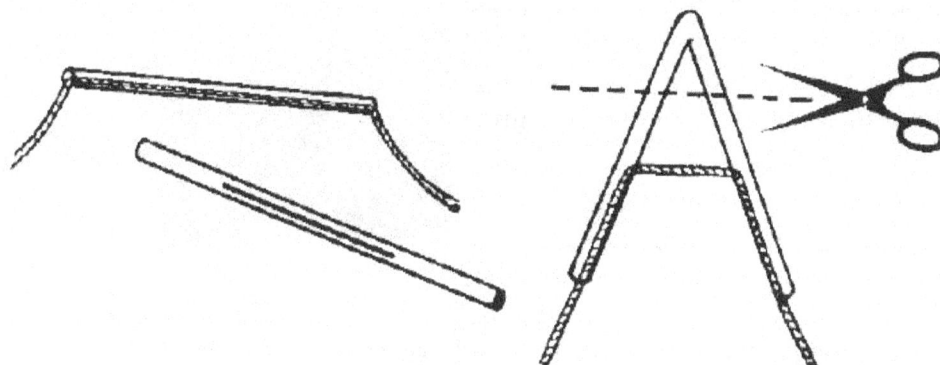

Doug Henning

Canadian magician, Doug Henning, is as well known for his flamboyant dress sense as he is for his magic. His brightly coloured clothes, thick moustache, long hair and warm smile made him instantly recognisable.

Performing magic, however, was not Doug Henning's initial chosen career - he wanted to become a doctor - his interest in magic was mainly a hobby. But, before starting medical school he thought about making some money performing magic. He persuaded a Canadian government panel that, as magic is an art form, he was eligible for a grant to study with some of the great magicians of the day.

This was the kick-start Doug Henning needed. Along with a friend of his, he assembled a rock and magic musical called Spellbound. Setting box office records, they renamed it The Magic Show before opening on Broadway and running, successfully, for over 4 years.

Doug Henning's World of Magic was the start of his TV career after the NBC television network liked what they saw in The Magic Show. This annual event ran for seven years, watched by millions of people each time.

One of Doug Henning's most famous illusions was the Double Sawing. Two women arrive on stage wearing different coloured dresses, one all in pink and one all in blue. They each enter their own box and are then sawn in half. The great twist here is that the bottom half of one women is put together again to the top half of the other women and vice-versa. When they emerge from their boxes, one has a pink top half but a blue bottom half and vice-versa. Incredible!

There was also Doug Henning's Miss-Made Girl, in which a young woman

stood in a vertical box of four equal heights of different colours. Blades were inserted between each one before the four cubes were then shuffled. When the fronts were opened up, the girl had her thighs on the top, her midriff next, followed by her head and finally her feet. The boxes were then rearranged and opened up for the girl to step out complete.

As well as great stage illusions, Doug Henning was also very good at close-up magic tricks. One of these was the Fickle Nickel where a nickel coin disappears and reappears in the palm of his hand even though he shows both sides of his hand whilst performing the trick.

In 1987, Doug Henning decided to leave magic behind and concentrate on Transcendental Meditation. It was rumoured that he was to come back into magic in late 1999 but, sadly, he died of liver cancer in early 2000.

42

Mysteriously Multiplied Money

Effect: Three coins increase to six in your hand.

Secret: Show three coins and count them out on to a magazine. Unknown to your audience, there are three more coins hidden inside the magazine cover. Fold the magazine slightly and tip the three visible coins into your hand. At the same time, the three hidden coins will slip from the magazine into your hand too but the audiences do not know about this. Wave your other hand over the one holding the coins. Slowly open your hand to show that the coins have mysteriously doubled.

3 COINS HIDDEN
UNDER COVER

Card Code

Effect: You identify a chosen card.

Secret: Place nine cards on the table. The suits do not matter but the values must be the Ace to the Nine. While you are out of the room, someone chooses any one of the nine cards. When you return, you know immediately which card has been chosen because you have a secret assistant who holds a pack with a nine showing. They secretly indicate the value of the chosen card to you by the position of his/her thumb over the nine pips. For this trick, you should be very tactful.

A Rising Match

Effect: A match rises from a matchbox.

Secret: Make a small hole on the top of a matchbox. Push a match into the hole as shown in the illustration. When you push the drawer, the match will be forced out through the hole. If you tilt the matchbox slightly with the end of the drawer facing the audience, it will give the impression that the match rises from the drawer as it is opened. Interesting, isn't it?

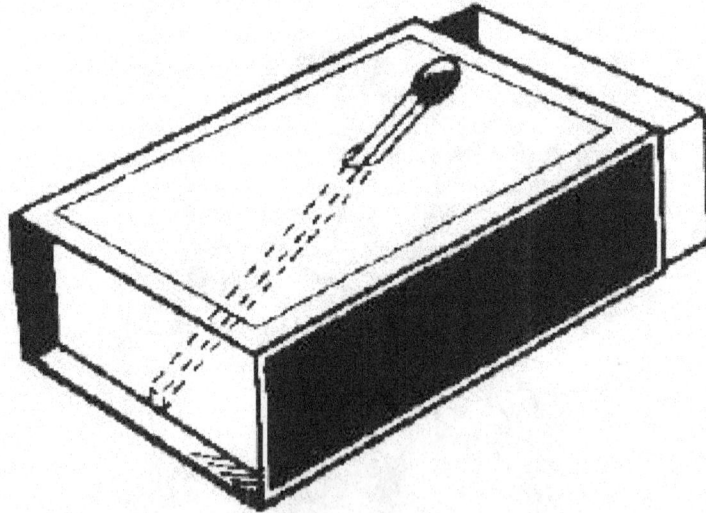

A Magically Changing Card

Effect: A playing card changes value magically.

Secret: You need an Ace of Clubs, a Queen of Hearts and one other card. Score across the centre of the Ace and the Queen and glue them together for half of their length. Use the third to form a back to the special card. Hold the card in your left

it?

THE THIRD CARD GLUED

A Chain Reaction

Effect: Paper clips form themselves into a chain.

Secret: Link 10 paper clips and put them in one corner of an envelope. Put a strip of glue on the inside of the envelope, just above the chain. Put 10 loose paper clips into the envelope and you are ready to show the trick. Tip out the loose clips and then drop them, one at a time, back into the envelope and seal it down. Tear off the corner of the envelope and the chain will drop out. Casually put the envelope in your pocket and display the chain of clips. For this trick, you have to be real quick and careful that the audience do not see through the envelope.

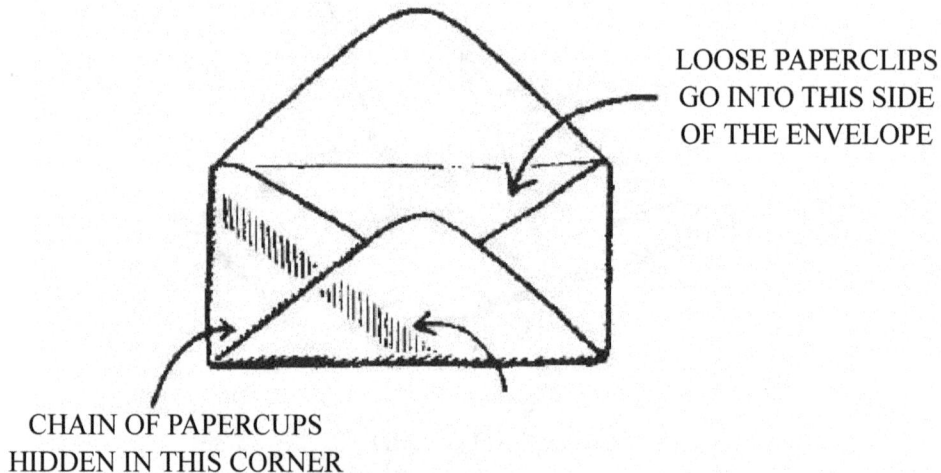

LOOSE PAPERCLIPS
GO INTO THIS SIDE
OF THE ENVELOPE

CHAIN OF PAPERCUPS
HIDDEN IN THIS CORNER

Penn and Teller

Penn and Teller are a comedic double act in magic who are somewhat eccentric in their performances. Teller, now legally his full name but originally Raymond Teller, does not speak during their performances so it is left to Penn Jillette to promote the illusion with his storytelling.

Penn and Teller met in 1975. It wasn't until 1981 that Penn and Teller performed as a duo only and just four years later, they appeared on television for the first time in *Penn and Teller Go Public*.

The tricks of Penn and Teller are often presented with a humorous slant, but also rely greatly on shock appeal with some of the gory effects they produce.

Not surprisingly, Penn and Teller have enraged other magicians for publicly revealing how some tricks are done. However, the tricks revealed by Penn and Teller are usually variants of other tricks we already know but they have stamped their mark on them, and the reveals are only what they have added to the original tricks.

Having said that, some old-school magicians still believe that revealing any magic trick to the outside community is against the principles of the magician's code.

But upsetting magicians has not stopped the rise of Penn and Teller. Throughout the 1990's, they managed to do several television one-off programmes and even a series for Channel 4 in the UK.

Both Penn and Teller have written quite a few books on the subject of magic, some they have co-written and others are separate pieces of work. Notable ones are 'How To Play In Traffic', 'How To Play With Your Food' and 'Cruel Tricks for Dear Friends'.

Indeed, Penn and Teller have hit the big heights of Las Vegas since 2001 where they perform six nights a week. Penn and Teller have also made it to Hollywood with several films under their belts.

Salt Cellar Suspension

Effect:

Secret: All you need, apart from the salt cellar, is a toothpick. Keep the toothpick

into the top of the salt cellar. Hold the hidden toothpick between your thumb and

hand the cellar to someone for examination, secretly drop the toothpick on your lap. Be very tactful while performing this trick.

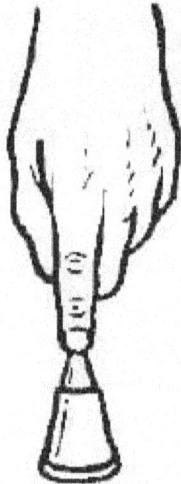

CONCEALED
TOOTHPICK

AUDIENCE VIEW

YOUR VIEW

A Cool Catch

Effect: Two pre-determined cards are caught in mid-air.

Secret: Secretly place the Seven of Clubs on the bottom of the pack and the Nine of Spades on the top. When you do the trick, remove the Seven of Spades and the Nine

pack in the air but keep a tight hold of the top and the bottom cards. Pretend you have caught the two you now hold. Show them triumphantly and everyone will believe that they are the two cards you placed in the centre of the pack earlier.

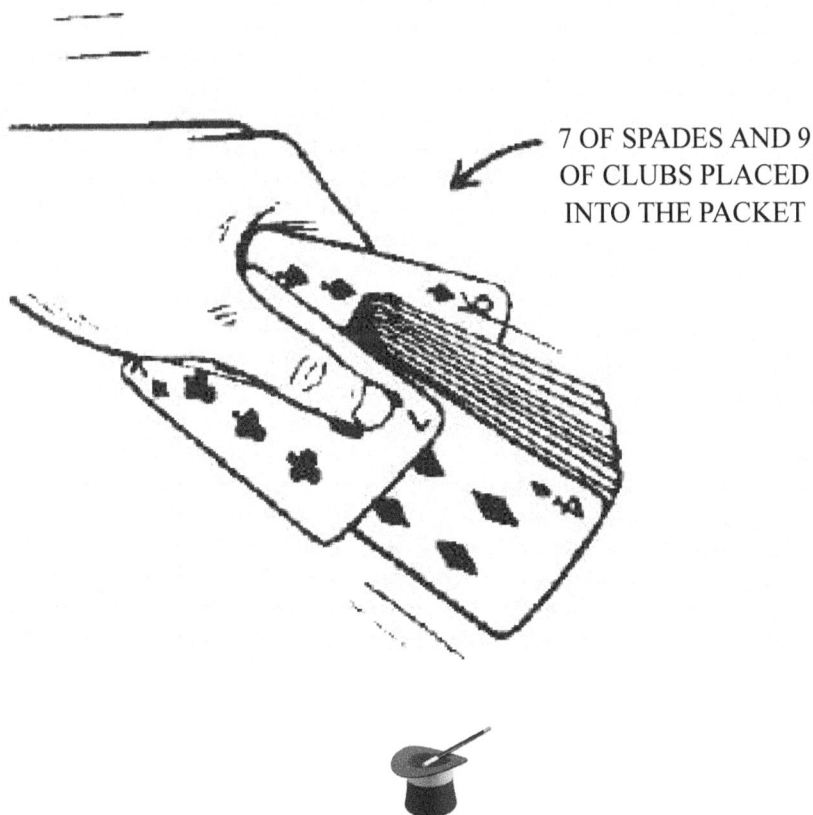

7 OF SPADES AND 9
OF CLUBS PLACED
INTO THE PACKET

United Handkerchiefs

Effect: Two handkerchiefs are thrown in the air where they knot themselves together.

Secret: Apart from the handkerchiefs, all you need for this trick is a small elastic band hidden in your right hand. Pick up one of the handkerchiefs with your left hand and place it in the right. As you do so, push a corner of the handkerchief into the elastic band. Do exactly the same with the second handkerchief. Throw the two handkerchiefs into the air and catch one of them as they descend. The elastic band holds the two handkerchiefs together and they appear to be knotted. This is simple, yet a clever trick!

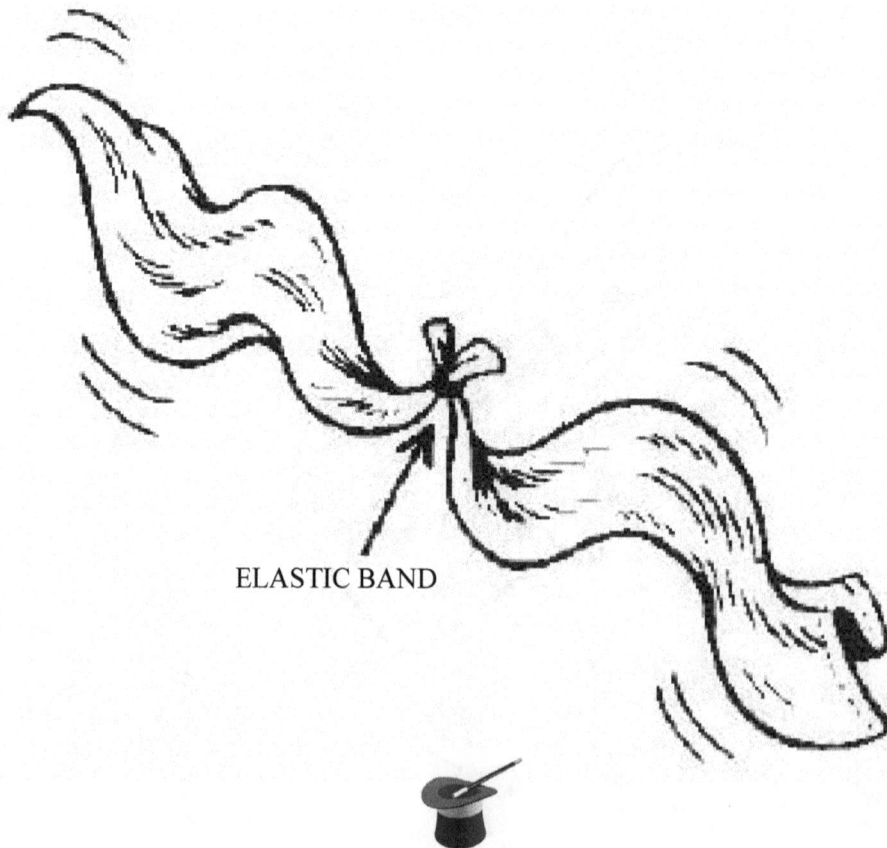

ELASTIC BAND

A Magical Sum

Effect: You predict the total of a sum.

Secret: Write 1089 on a piece of paper, fold it and then place it on your table. Ask someone to choose any three-digit number. All the three digits must be different

Open the paper – your prediction is the same as the spectator's total. In fact, if the calculations are done correctly, the answer is always 1089! For this, you have to use your mathematical skills, as well!

$$
\begin{array}{rl}
9\ 8\ 3 & \leftarrow \text{ANY NUMBER} \\
-\ 3\ 8\ 9 & \leftarrow \text{REVERSED} \\
=\ 5\ 9\ 4 & \leftarrow \text{SUBTRACTED TOTAL} \\
+\ 4\ 9\ 5 & \leftarrow \text{REVERSED} \\
=\ 1\ 0\ 8\ 9 & \leftarrow \text{ADDED TOTAL}
\end{array}
$$

THE TOTAL MATCHES
YOUR PREDICTION!

51

A Rope through the Body

Effect: A rope passes through your assistant's body.

Secret: Two spectators are invited to examine a long length of soft rope. Put the middle of the rope over your assistant's head to his/her front. Take both ends of the rope around the back of his/her body where you apparently cross over the ends. In fact, your assistant secretly sticks up his/her thumbs and you wrap the rope around them, as shown. Hand the ends of the rope to the spectators on either side. When you clap your hands, your assistant secretly sticks up his/her thumbs and walks forward – free of the rope.

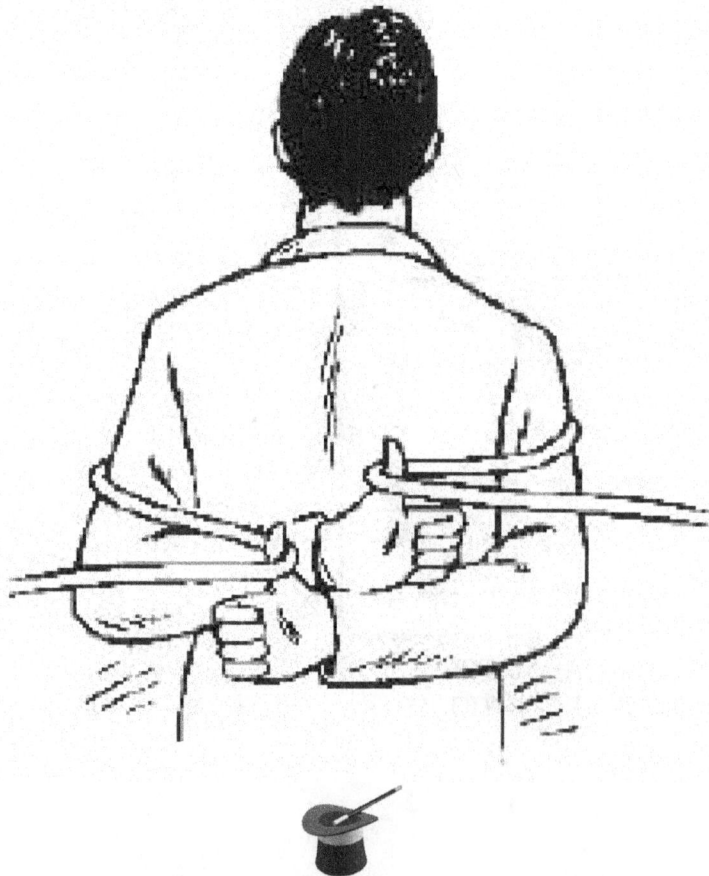

Siegfried and Roy

SARMOTI - A magical word used in the spectacular stage shows of Siegfried and Roy. What does it mean? Siegfried and Roy Masters – of the Impossible.

A word which evokes passion and commitment to Siegfried and Roy's superb performances, encompassing the animal instinct when thought of with the white tigers and white lions which are very much a part of the magic of Siegfried and Roy.

Born as Siegfried Fischbacher and Roy Horn in Germany during the Second World War, it was Siegfried whose interest in magic started at an early age. Roy was more interested in animals, often visiting Bremen Zoo, Germany, where he befriended a cheetah named Chico.

It was only when they happened to both take jobs on the ocean liner TS Bremen did Siegfried and Roy meet and become a double-act, along with Chico the cheetah. After five years working together on the ship, they decided to take their act to the masses, working in Germany and touring Switzerland and Monte Carlo.

In 1967, Siegfried and Roy made their Las Vegas debuts starring at the Tropicana before going back to Europe. Three years later, in 1970, they were back in Vegas, where Sahra, a Siberian white tiger became part of the act.

After working in Puerto Rico and Vegas during this decade, they knew they had finally made it really big in Las Vegas during 1978 with February 17 being declared Las Vegas Siegfried and Roy day.

With Siegfried and Roy's love of the big cats, it was the Maharajah of Baroda who asked them to help to save the white tiger. They received three cubs who went on to successfully produce offsprings.

The magic and illusions of Siegfried and Roy are very much old school in the type of tricks they perform, but they manage to produce a spectacular result with their versions on stage. In particular, the squeeze box illusion, where an able assistant enters a long box with her head and feet showing. This is then squeezed at both ends so she is only, say, 12 inches tall. The box is then lengthened again and out she pops, normal height again. All this is done on a see-through table in under a minute.

Siegfried and Roy's illusions led them to achieve the ultimate accolade of "Magicians of the Century" in the Millennium year 2000.

Disastrously, in October 2003, Roy was bitten by one of his tigers during a performance, after stumbling on stage. He also suffered several strokes and underwent surgery on his windpipe. He had to undergo many months of treatment and made excellent progress which enabled him, along with Siegfried, to make a final stage appearance in February 2009 at *The Bellagio*, where they performed their *trademark illusions*.

A Soapy Secret

Effect: A coin is shown on the palm of the left hand. A moment later the coin has vanished.

Secret: This trick only works with a small coin. The coin is shown on the palm of

hand it seems that the coin has disappeared – but it is really stuck to the back of the

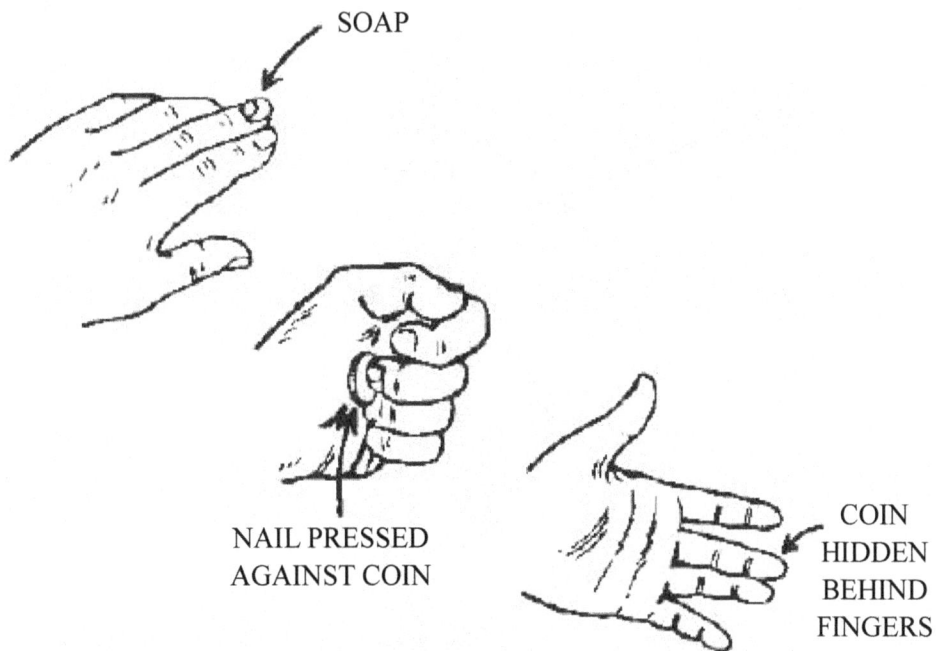

SOAP

NAIL PRESSED
AGAINST COIN

COIN
HIDDEN
BEHIND
FINGERS

A Correct Prediction

Effect: You predict the numbers on two dice.

Secret:
envelope. Glue two dice in one end of the drawer of a large matchbox with the four

the matchbox to show the loose dice. Close the drawer and ask a spectator to shake the box. Take the box back and open the drawer at the end containing the glued dice. Allow someone to see the dice and then close the drawer. Your prediction is opened and is, of course, correct.

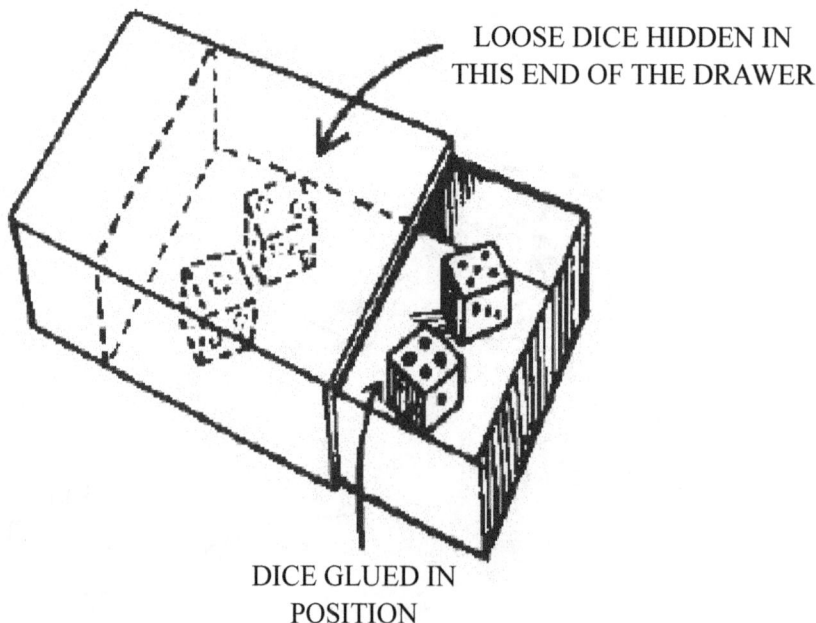

LOOSE DICE HIDDEN IN
THIS END OF THE DRAWER

DICE GLUED IN
POSITION

Bell Rings on Its Own

Effect: A small bell rings of its own accord.

Secret: A small bell is on your table. Hold an opaque scarf in front of you, between your hands. In the top right corner of the scarf, there is a small pin. Show the other side of the scarf by moving the right hand to the left and the left hand to the right. Do this a few times, and then pin the scarf to your left shoulder. This leaves your right hand free to reach down and ring the bell. Immediately bring your hand back up to recover the scarf from your shoulder. However, you have to be really quick with your hands and the scarf to perform this magic!

PIN

Clip The Card

Effect:
fails and the clip jumps to the end card!

Secret:
from the rest. Show the cards to a spectator. Hand him the clip and ask to put it on the centre card when you have turned the cards face down. He puts the clip on the centre card. When you turn the cards face up, the clip is actually on the end card! The trick works by itself – but there is no need to tell your audience about that!

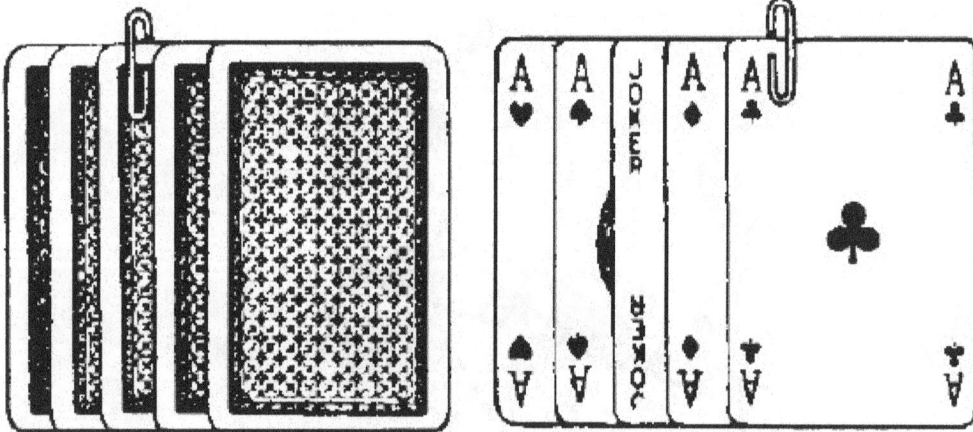

5 CARDS GLUED TOGETHER

The Wandering Coin

Effect: A coin travels from one hand to the other.

Secret:

right hand, as shown. Turn over your hands quickly and simultaneously. Lift the right hand and the coin has vanished. Lift the left hand to reveal that there are two coins beneath it. Due to the positioning of the coins, this trick works automatically. Even so, you should still practise it in private before showing it to anyone. If you

is reversed.

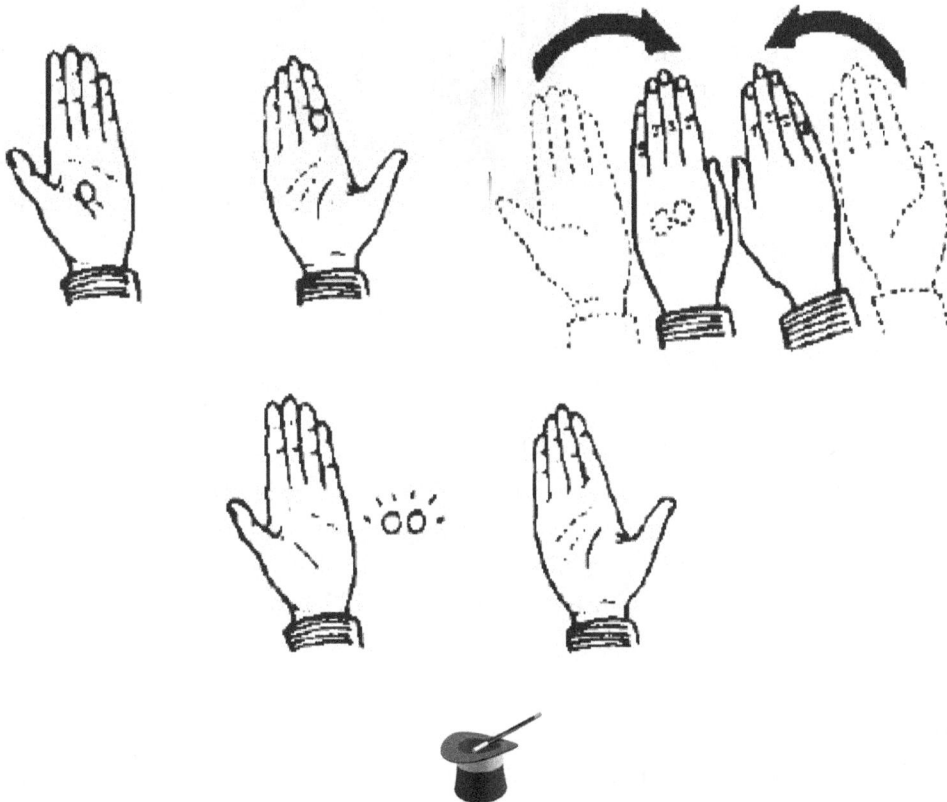

Lance Burton

Lance Burton is one of the best sleight-of-hand magicians and certainly deserves the title of 'A Magician's Magician'.

When he was just five years old, he was at a party with a show being done by a magician and was a volunteer to one of the magician's tricks. After being given a book by a neighbour, he quickly learned the ten magic tricks it contained and did his first performance for the local children where he lived. Lance Burton was hooked.

In his mid-teens, Lance Burton entered his first competition as a junior magician and easily took first prize. By now, the very magician who had mesmerised Lance Burton all those years ago had become his mentor.

After moving to Southern California, he managed to get a spot on Johnny Carson's The Tonight Show. This catapulted the name of Lance Burton into the homes of the American public. From that moment, offers of work began to flood in.

After starting an eight-week stint at the Folies Bergere show in Las Vegas, Lance Burton was still there an amazing nine years later. He also wanted to compete at the International Federation of Magic Societies main event - the World Championship of Magic. He won for America in 1982 - indeed, Lance Burton was the first American to do so.

It was this Championship act that featured the great sleight-of-hand of Lance Burton to produce things from nowhere.

A dove appears from his hand which only moments before was on fire and next, he takes off his gloves, puts them together, folds them inside out and seems

to throw them away, only for another dove to appear with the gloves nowhere to be seen.

A silk scarf is shown, held at both ends, and his hands are brought together. When his hands are apart again, a long candle is revealed which, when lit, just as quickly disappears.

Lance Burton now has a theatre named after him in Las Vegas where he performs nightly to packed houses. He has done several TV specials and has become a regular guest on major talk shows.

Long may the legend that is Lance Burton continue!

Food and Drink

Effect: From cards bearing the names of foods, you pick out the odd card on which is written the name of a drink.

Secret: You need six blank cards, one of which is little longer than the others. Hand

kinds of foods. The long card is handed to someone with the request that he writes the name of a drink. The cards are mixed up and handed back to you. Without looking, you immediately pick out the drink card. You can detect it quite easily by touch, but be tactful enough that nobody sees through your trick.

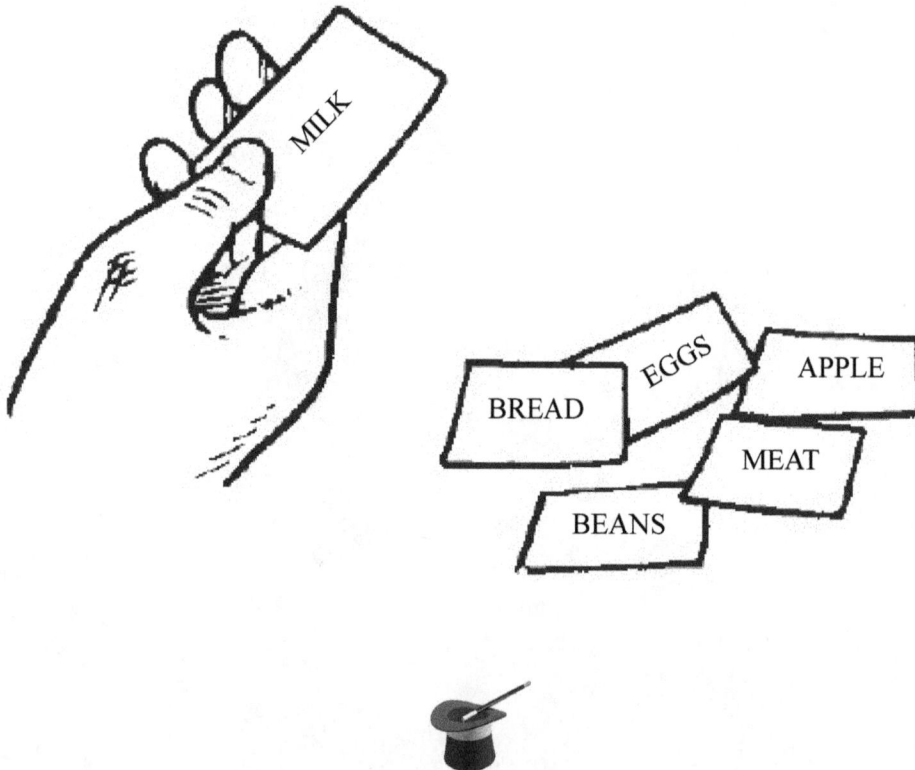

Discovery of Aces

Effect:

Secret: Secretly remove the Aces from the pack. Push them into a paper clip which is pinned to the inside of your coat at the rear. When you want to show the trick,

them behind your back and say you are about to attempt the impossible. Secretly remove the Aces from the hidden clip and bring them forward one at a time as if you have searched through the pack for them.

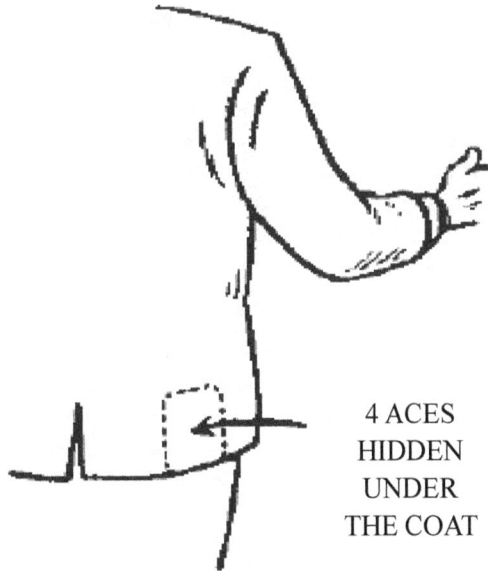

4 ACES
HIDDEN
UNDER
THE COAT

An Unburstable Balloon

Effect:
the balloon does not burst.

Secret: The balloon has on it several pieces of clear or transparents adhesive tape. These will not be visible from even a short distance. Push a pin into the balloon at

ability. When you want the balloon to burst, simply push the pin into an area that is not protected by the tape. This is real smart isn't it?

Spin Vanish

Effect: A matchbox is banged on a spinning coin to stop it. When the matchbox is lifted, the coin has vanished.

Secret: Spin a small coin on the table top. Hold the matchbox, which must be empty, so that the drawer has its open side down. Bang the matchbox sharply on top of the coin. Ask a spectator if he thinks the head or tail side of the coin is uppermost. Lift the matchbox and the coin has disappeared. Because you banged the matchbox down so hard, the coin penetrated the cover and is now inside the box. Be careful that you hold the mathcbox correctly with the drawer, opening down wards.

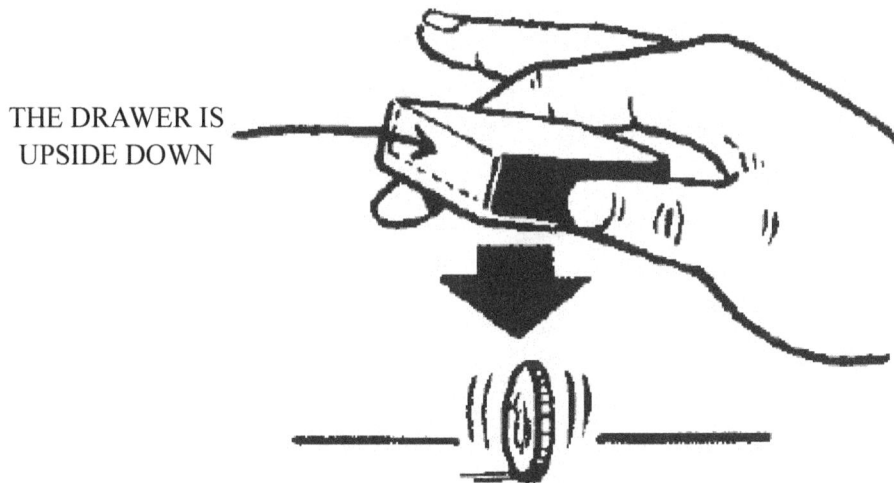

THE DRAWER IS
UPSIDE DOWN

Money, Money, Honey!

Effect: Spectators try to win some money – but you win.

Secret:

of paper. You have secretly marked the money envelope with a pencil dot. Ask someone to mix the envelopes and then return them to you.

Move the marked envelope to the second position. Ask someone to spell, "money", transferring one envelope to the bottom of the pile for each letter. He keeps the envelope that falls on the letter, Y. Do the same with three other spectators. You are left with one envelope, the one containing the money.

THE DOT IDENTIFIES THE
ENVELOPE CONTAINING ₹ 5 NOTE

Where are the Matches?

Effect: Two empty matchboxes and one containing some matches are moved around on a table. The spectators are unable to identify which box contains the matches.

Secret: All three matchboxes are empty. Attached to your right arm and hidden by your sleeve is a matchbox containing some matches. Shake a box with your left hand. It sounds empty. When you mix up the three visible boxes, you can make any one appear to be the full one, simply by shaking it with your right hand. The spectators think it contains matches because they hear the sound from the hidden box. Be careful that nobody sees or watches your right arm as you move it quickly to perform the trick.

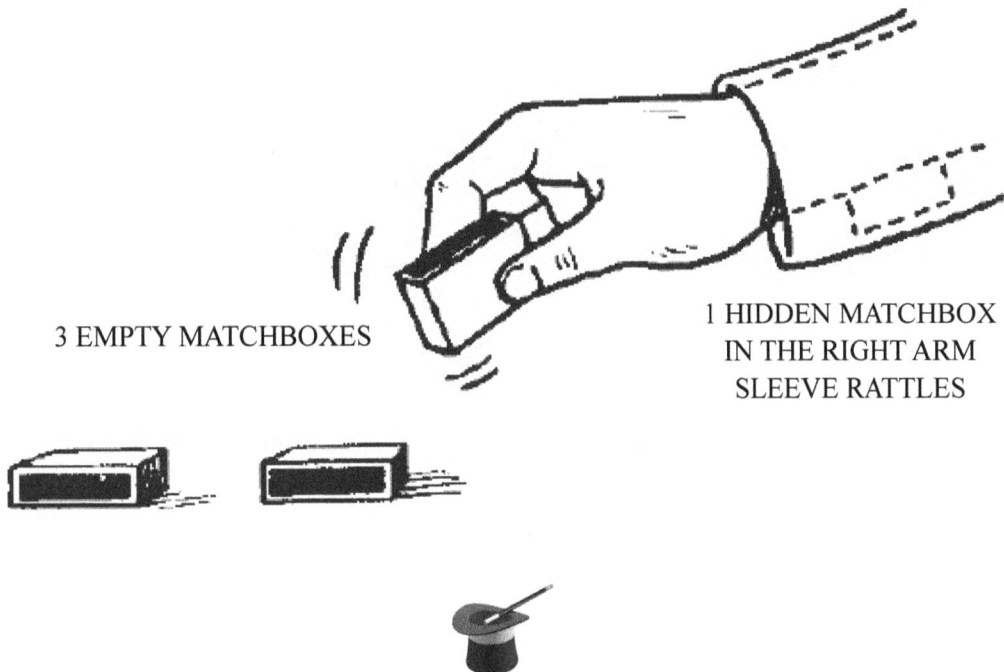

3 EMPTY MATCHBOXES

1 HIDDEN MATCHBOX
IN THE RIGHT ARM
SLEEVE RATTLES

Anti-Gravity Cards

Effect: Several playing cards adhere to your hand, until you tell them to fall off.

Secret: You need to make a special card. It is simply an ordinary card with a tab

between the special card and your hand, as shown. Turn your hand over and the cards will appear to be sticking to it – until you let go of the tab. This needs some of

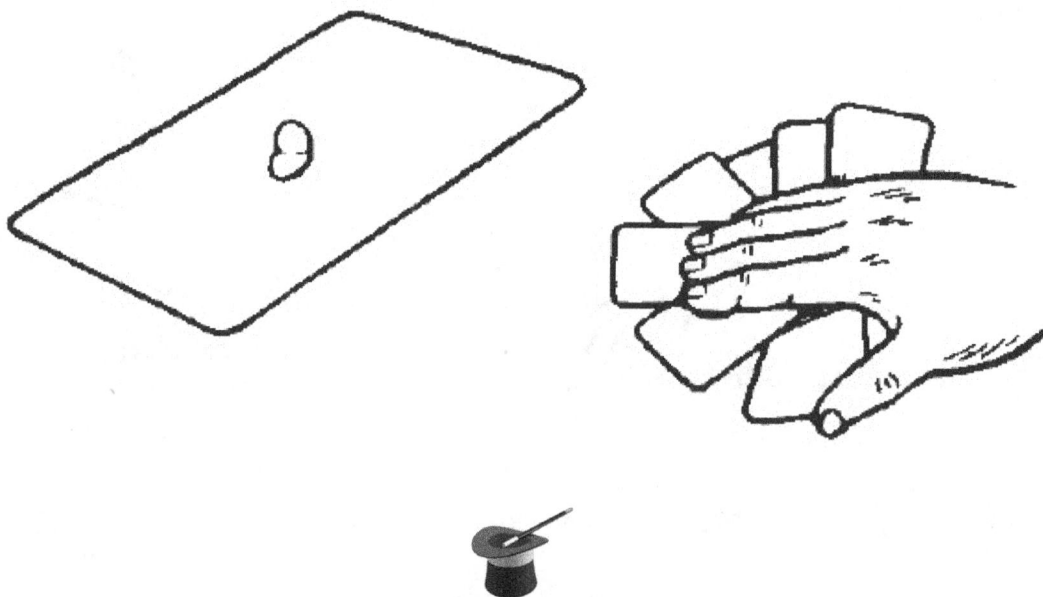

A Coin in the Glass

Effect: A coin disappears.

Secret: Put a coin in the centre of a handkerchief draped over your right hand. Place a glass tumbler on the palm of your left hand. Turn your right hand over, retaining a grip on the coin and bring the handkerchief over the glass. When the glass is

on your table and casually drop the coin in your pocket. Remove the handkerchief, and lo! The coin has vanished.

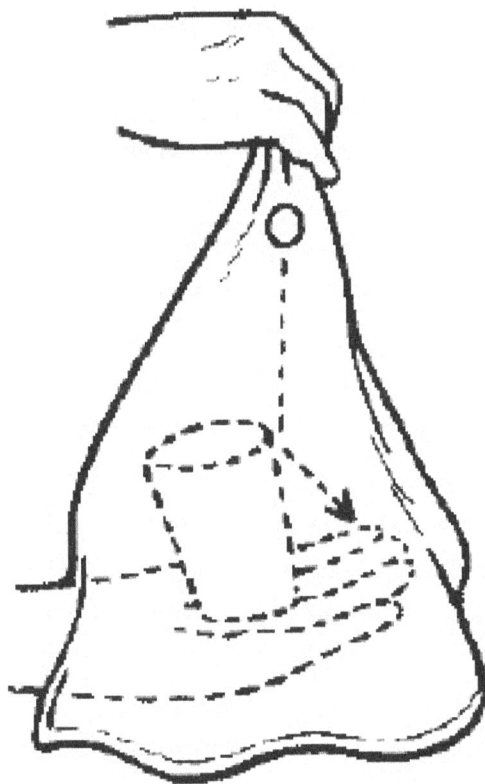

A COIN HITS THE GLASS AND DROPS ON TO THE LEFT FINGERS

Vazhakunnam Neelakandam Namboothiri

Vazhakunnam Neelakandam Namboothiri is also known as the *Grandfather of Magic* in Kerala. He gave a new meaning to magic in Kerala. He played an important role in bringing Magic as an Art. For more than half a century, he had kept the onlookers spellbound in spite of his extreme simplicity, and devoid of any elaborate stage and costume.

He was born on Makeeram star on 26 Makaram, 1078 (08-02-1903) in Vaazhakunnath Mana in Thiruvegappura, After committing Vedam to memory, he learned Sanskrit at Guruvayur from his brother, Vasudevan Nambudiri (scholar and expert on discourse), and English from Pattambi Narayana Iyer.

While there, he happened to watch "Cheppadividya" [tricks using Cheppu (cups) and Panthu (balls)] peformed by Pallatheri Nambyathan Namboodiri an expert and his interest in magic was rekindled.

Vaazhakunnam later became famous for Kayyothukkam, although occasionally he performed also "Cheppum Panthum" (cups and balls) to small family gatherings. Gradually, he mastered other modern magic tricks also. He learned "bullet" tricks from the renowned magician Bekkar of Alappuzha and taught him *Cheppum Panthum* in return - as *Guru Dakshina*. He became adept in card tricks from Alli Kittan (Alli Krishna Iyer). Once in a while, depending on the audience, he would show the "bullet", "tying upon the cross", "guillotine" and other technology-oriented tricks. Though he did not consider hypnotism and mesmerism as part of pure magic, there were occasions when he had to do such tricks under pressure. He was famous for his disappearing act (Mooti Vidya). There were many instances where, after performing a trick, he would even show how it is done, saying that with practice, anyone could do it; such was his humility.

It was after 1940 that he started real stage performances with his troupe. Apart from magic, the shows included short dance programmes, comedy skits, etc. With him was Paryanampatta (Kunchunny Nambudiripad) who was already a well-known actor and Vaazhakunnam's disciple in magic. After 1948, he modified his shows including costumes. He would start with a two-line prayer (his own creation) and a Gandhi Sthuthi, being a true Gandhian. He would start from very simple tricks and proceed to more complicated ones. He was unmatched in "producing" any number and almost any size of different articles apparently from nothing.

At some point, he and P C Sircar (Jr) met, and ever since they had tremendous mutual respect. In his old age, Sircar visited him at home and gave him a cycle rickshaw and in return received a sword. Most later magicians considered it fortunate to be Vaazhakunnam's student. Apart from Paryanampatta, those trained by him include the late Manjeri Ali Khan, Prof. Muthukad, R K Malayath, Joy Oliver, K P Krishnan Bhattathiripad, Kuttiyadi Nanu, K S Manoharan, K J Nair, Vadakkeppad Parameswaran, Raghavan, his own sons and others.

He had married K C Anujathi Thampuratty of Kottakkal Kovilakam in 1980. They have two sons and a daughter. The only time he could not keep an appointment on stage was in 1983 at Kasargod, but that was because on the February 9, that year, he disappeared forever from the magic (and real) world!

Colour Sense

Effect: You identify the colour of a crayon handed to you behind your back.

Secret: Ask someone to take any crayon from a box and hand it to you behind your back. You then turn to face the audience but keep the crayon behind your back. Secretly dig your right thumb into the crayon. Keep hold of the crayon behind your back as you bring your right hand up to your forehead as if concentrating.

Take a quick look at your nail – you will see bits of crayon in it and will know the chosen colour.

Concentrate some more and then announce the colour.

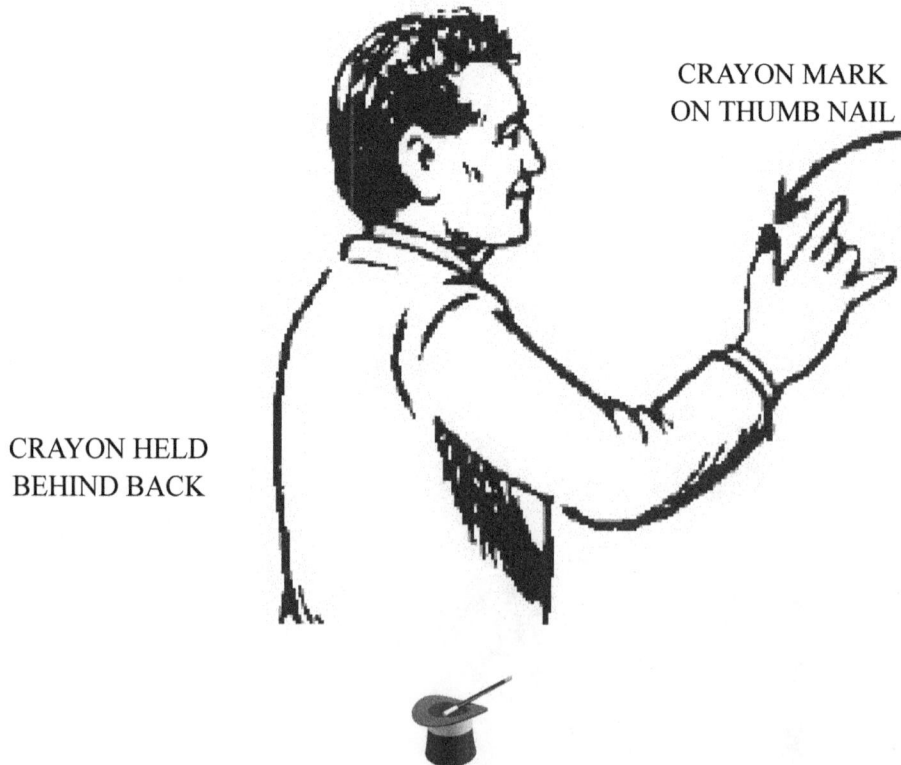

CRAYON MARK
ON THUMB NAIL

CRAYON HELD
BEHIND BACK

Matches Defy Laws of Gravity

Effect: A box of matches is opened and the drawer held upside down. The matches defy the laws of gravity and do fall out – until you command them to.

Secret: Wedge a match across the drawer of a box of matches. You will probably have to shorten the match to do this. When you take the drawer from the box, hold it by the sides to keep the secretly-wedged match in position. Command the matches to fall and release the pressure on the wedged match. All the matches fall from the box on your command. Remarkable, isn't it?

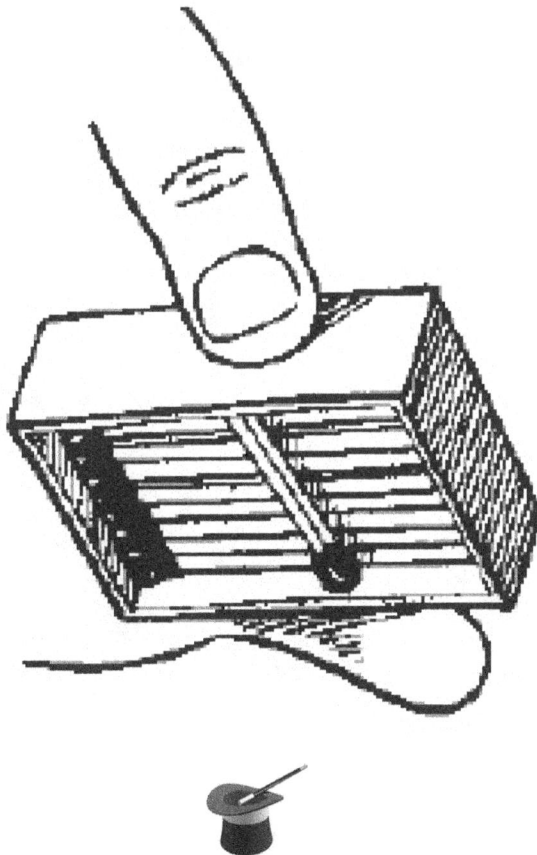

67

The Haunted Pack

Effect: A pack of cards divides to reveal a chosen card.

Secret: Ask someone to take a card, look at it and remember it. Point to the top card of the pack. As you do this, you secretly drop a few grains of salt on the top card. The spectator puts his card on the top of the pack and the pack is then cut, so the chosen card is hidden somewhere near the centre. Hit the edge of the pack with the heel of your hand and, thanks to the secret salt, the pack will divide at the chosen

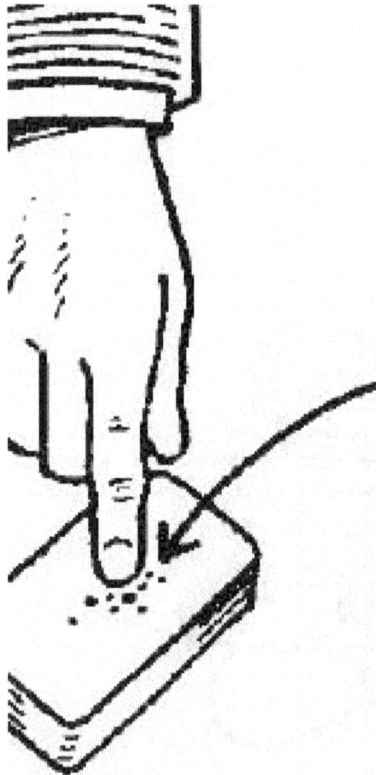

FEW GRAINS OF SALT
DROPPED ON THE TOP
CARD AS YOU POINT TO IT

Two in the Hand

Effect: Someone thinks of one of the two items, you predict it correctly.

Secret: Give someone any two items. They could be coins, cards, or buttons – anything you like. He holds one in one hand and the another in the other. While your back is turned, ask him to think of either object – he has a free choice. To help him concentrate, suggest that he holds the chosen object up to his forehead. After a short while, ask him to lower his hand and you then turn around. You know which object he chose as the hand holding it will be lighter than the other. Holding it up

magical one!

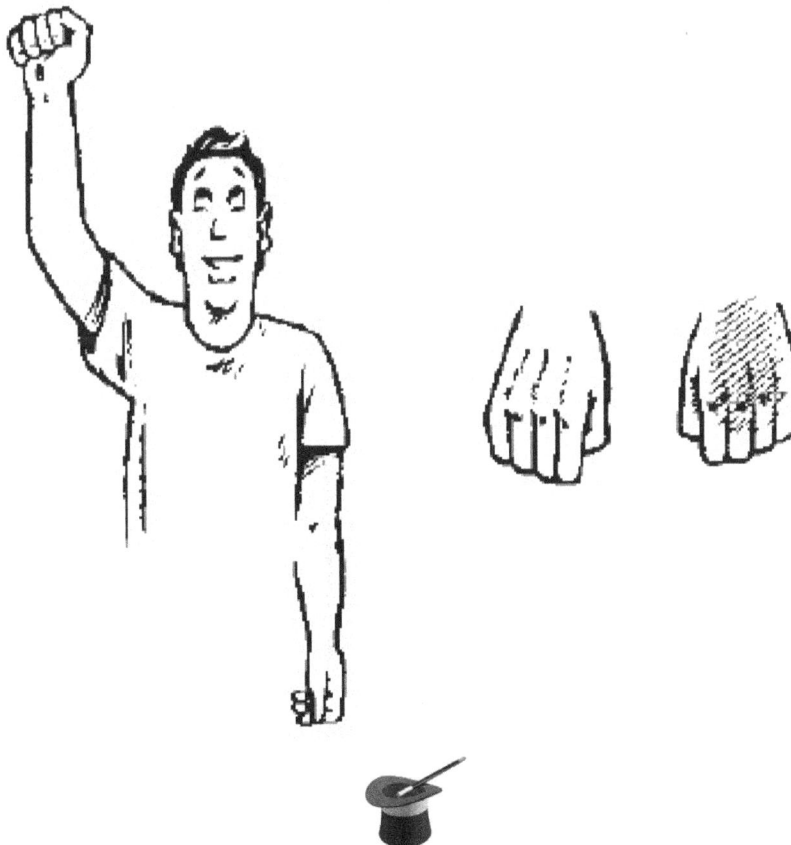

Match Restoration

Effect: A matchstick is broken into half and then magically restored to whole one.

Secret: You need a handkerchief with a matchstick hidden in its hem. A match is taken from a matchbox and wrapped in the handkerchief. Ask a spectator to break the match through the material but make sure it is the one hidden in the hem that he breaks. Shake out the handkerchief and the match just removed from the box falls

back into your pocket. It appears that the match was restored by magic. This is a quick one so be sure that the broken matchstick hidden in the hem goes back into your pocket and doesnot fall down.

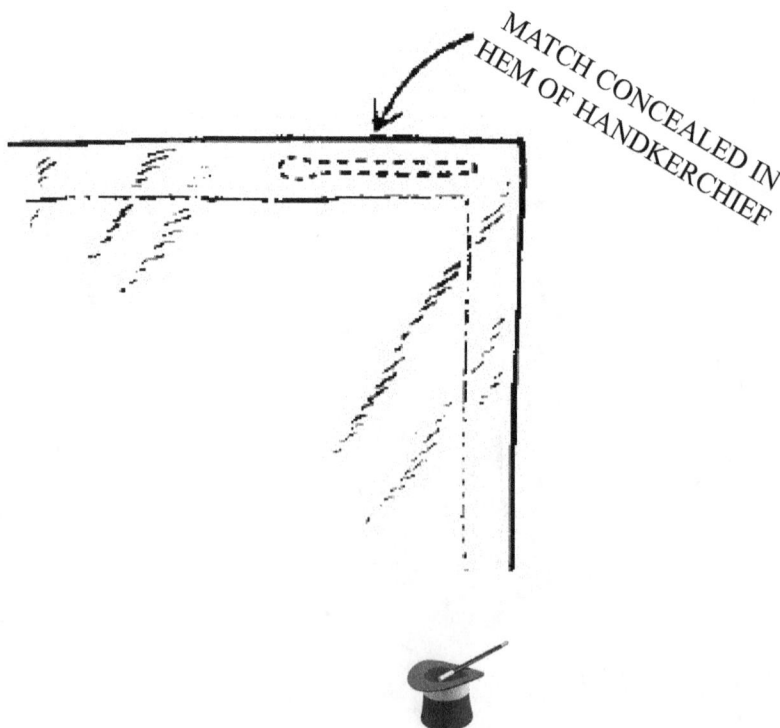

MATCH CONCEALED IN HEM OF HANDKERCHIEF

A Disappearing Ring

Effect:

Secret: The ring is tied to a long piece of elastic which runs up your sleeve to a safety pin attached to the top of the sleeve. The elastic should be long enough to

the air. When you let go, the elastic carries the ring up your sleeve. You could, if you

start of the trick.

Note: However, this trick needs practice and quick hand movements.

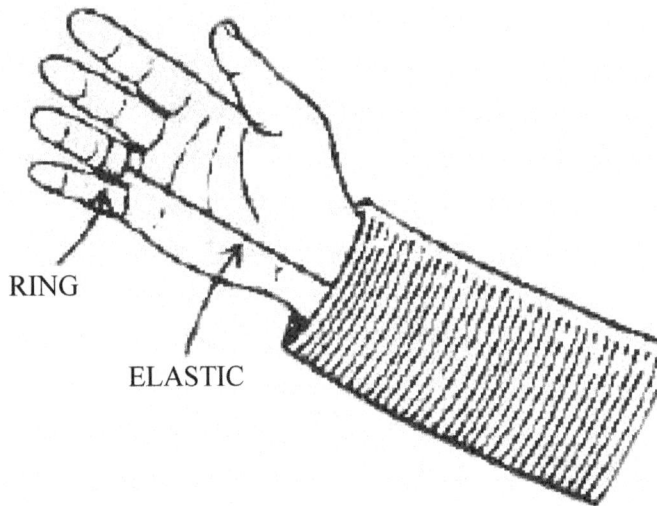

RING

ELASTIC

Spirit Initials

Effect: A spectator's initials appear on his own hand.

Secret: Hand someone a soft pencil and ask him to print his initials on a sugar cube. Take the sugar cube and secretly press your thumb against the initials. Drop the sugar in a glass of water. Ask the spectator to hold his hand over the glass. To show him what to do, you guide his hand over the glass. As you do this, press your thumb against his palm. You have now transferred a copy of the initials to his palm. He will be surprised when he sees the initials on his palm. Interesting, but be tactful while holding the spectator's hand to transfer a copy of the initials to his palm.

Locating the Coin

Effect: A coin is covered with any one of the three cups and the cups are then moved around. You can say immediately which cup is covering the coin.

Secret: Attached to the coin is a long hair. When the coin has been covered the three cups can be moved around as much as anyone wants. Although your back has been turned while this has been going on you can easily locate which cup hides the coin. All you have to do is look for the hair sticking out from beneath one of the cups and that tells you all you need to know!

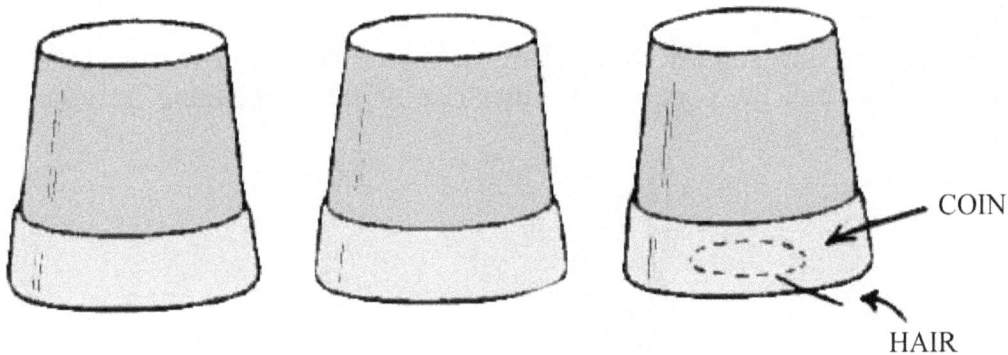

COIN

HAIR

2

Pencil in Command

Effect: The magician places a pencil into a bottle. It reacts to his commands and slowly rises and falls in the bottle.

Secret:

Props:

Preparation:
(18 inches to 24 inches in length) is tied, taped or applied with wax to one end of the pencil. The other end is tied to a button on the magician's clothing.

Presentation:
illustration). By moving slightly away from the table, the thread will become tight and on command, the pencil will either rise or fall, depending on your movements.

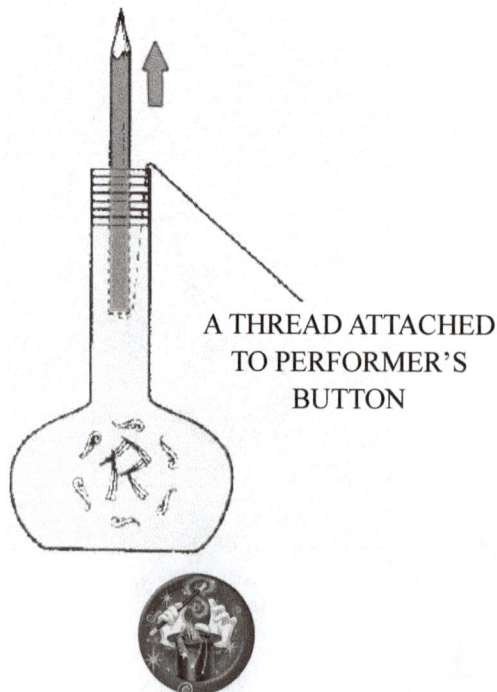

A THREAD ATTACHED TO PERFORMER'S BUTTON

Dice Deception

Effect: You read someone's mind.

Secret: Whilst your back is turned (or you can be blindfolded), someone throws a dice on the table. Ask him to remember the number thrown and then to make another

Ask him what total he has reached. Mentally subtract 25 from the answer he gives.

and the second is the second number thrown. Now pretend to read the spectator's mind and tell him what numbers he threw. Great, isn't it?

4

Changing Card to Matchbox

Effect: A playing card changes into a matchbox.

Secret: Trim a playing card and glue it to the top of a matchbox. Carefully fold the card over the box, as shown. Glue a matchbox label on the back of the card so that when the card is folded, it looks like an ordinary matchbox. Show your audience the playing card. The matchbox is hidden behind it. Bring your hand over the card and secretly fold it up over the box. Remove your hand and it seems that you have made a playing card change magically into a matchbox. However, you have to be quick and real smart while doing this trick!

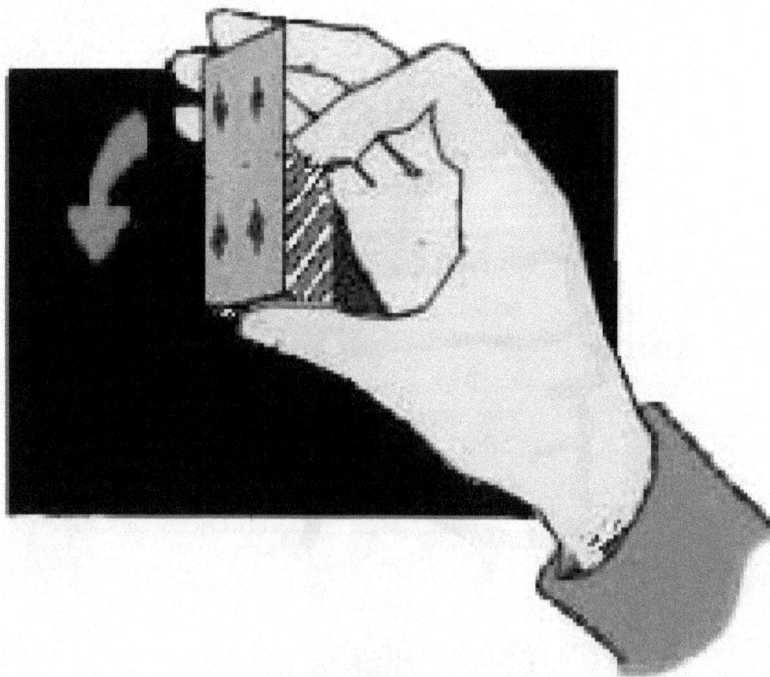

Making Money

Effect: Three coins are placed in your hand. When you open your hand again the

Secret: Before your performance, secretly stick two coins to the underside of your table with plasticine. Place three coins on the table. Let everyone see that your hands are empty. With your right hand, scoop the coins from the table into your

hidden coins. Open your left hand and the coins have magically increased. Wow, this is real smart!

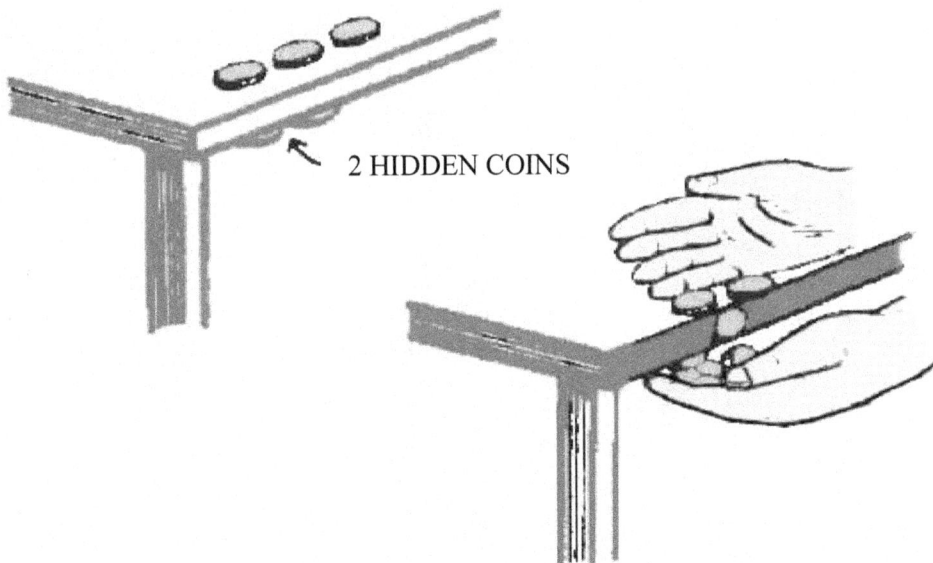

2 HIDDEN COINS

6

Cut and Restored Ribbon

Effect: A ribbon is cut in two and then restored.

Secret: You need a piece of ribbon, which is a metre long and another eight centimetres long. Sew the short piece to the middle of the long piece. Provided you keep it moving, the short piece will not be noticed. Fold the ribbon in half and

through the short piece and then cut away the rest of it (including the stitched bit). You have cut the ribbon in two, but you can now show it completely restored!

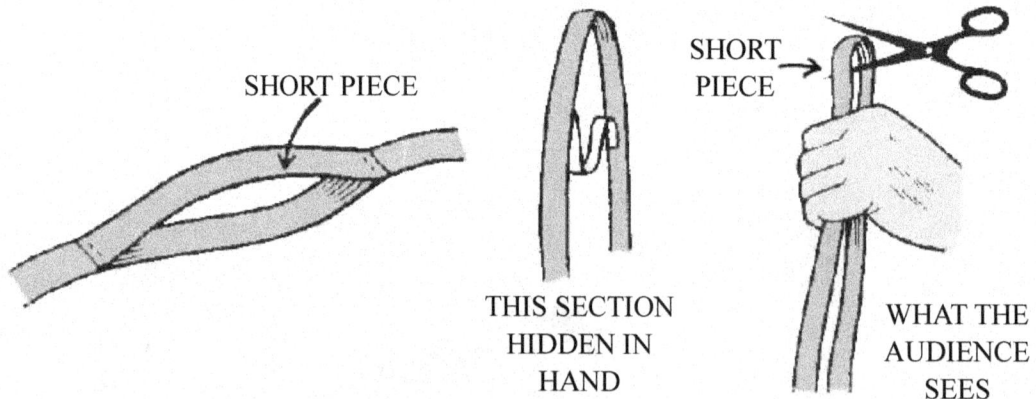

SHORT PIECE

THIS SECTION
HIDDEN IN
HAND

SHORT
PIECE

WHAT THE
AUDIENCE
SEES

Red or Black?

Effect: From a pack of cards spread face down on a table, you are able to tell which are red cards and which are black.

Secret: First separate the reds from the black. Bend all the red cards upwards. Bend all the black cards downwards. Mix the cards together. Spread all the cards face down on your table. Throw them down haphazardly and then invite a spectator to mix them up even more, so it is obvious that they are not in any special order. Ask the people to point to the different cards and you can tell them if they are red or black by the way they bend.

BLACK CARDS

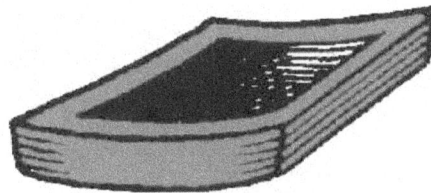

RED CARDS

8

Match Maker

Effect: Matches appear in an empty box.

Secret: The matchbox drawer is secretly prepared by cutting it in two. One section

matches. To show the box empty pull the long part of the divided drawer from the box. The matches will remain in position inside the cover and the drawer will appear to be empty. Close the drawer. To show the box full, push the drawer from the other end. This moves both the pieces of the drawer and the matches make their appearance. Wow! you have indeed created matches in an empty box.

MATCHES LEFT
BEHIND IN COVER

PUSH

A Vanishing Coin

Effect: A coin disappears from your hand in spite of the fact that several people check it is there.

Secret: Show the coin on the palm of your hand and cover it with a handkerchief. Go to several people and ask them to feel the coin under the handkerchief. The last person you approach is (unknown to the rest of the audience), your secret assistant. He pretends to feel the coin but actually takes it off palm. You can now pull the

coin has disappeared. This is a real smart one!

Anti-Gravity Ball

Effect:

Secret: The secret is a long loop of dark thread which is on your table with a ping-pong ball resting on it. Put your thumbs through each end of the loop. Lift your

rolling along the thread). You must wear dark clothing for this trick and be a little distance from your audience so that the thread cannot be seen.

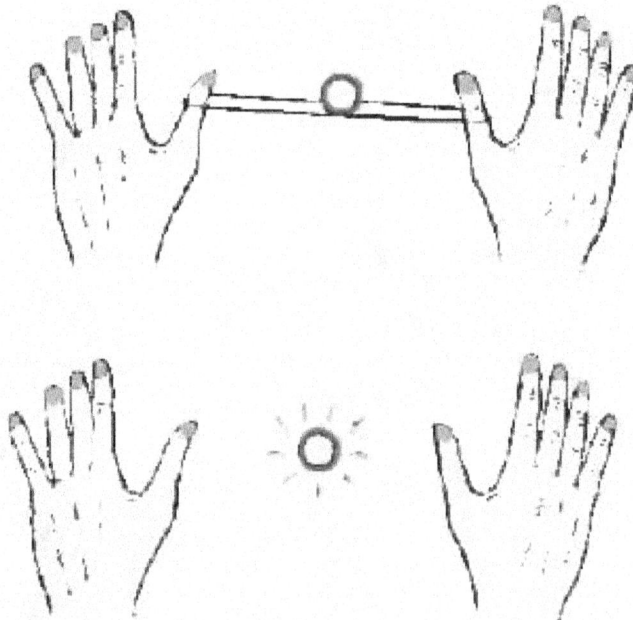

71 Science Experiments Series

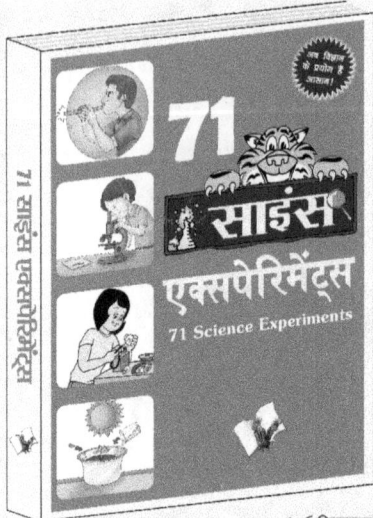

12140 P • ₹ 110 • 160 pp

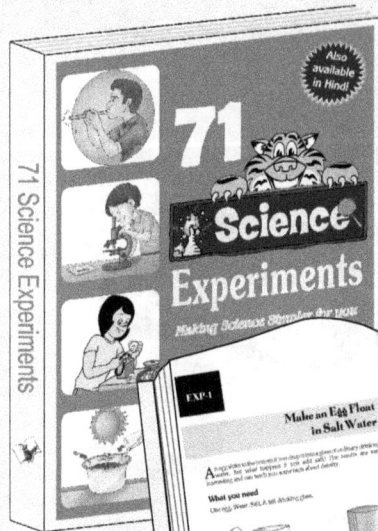

02139 P • ₹ 110
160 pp

A study of Science and Scientific theories and Laws is incomplete without relevant and methodical Experiments. In fact, Experiments are an inseparable part of any Scientific Study or Research. In this book, the author has tried to simplify science to the readers, particularly the school-going students through easy and interesting experiments. All the experiments given in the book are based on some scientific phenomena or another. Variation in atmospheric pressure, high and low temperatures, boiling, freezing and melting points of solids liquids and gases, gravitational force, magnetism, electricity, solubility of substances, etc. alters the experimental results. Thus, read each of these fun-filled experiment and carry it out in your homes or schools under the supervision and guidance of your teachers, parents or elders. The language used in the book is simple and all the experiments have been illustrated with relevant diagrams and methodical steps strictly based on scientific facts. So dear children, grab this book and satisfy your scientific curiosities by performing these incredible experiments to learn science with fun.